思科系列丛书

路由交换技术
项目化教程入门篇

袁劲松　胡建荣　主编

電子工業出版社
Publishing House of Electronics Industry
北京·BEIJING

内 容 简 介

本书基于 Cisco Packet Tracer 7.2.1 模拟器开发了由简到繁、从单项到综合、层次递进、环环相扣的 20 个实训项目，内容包括登录路由器、密码恢复和基本配置、配置文件和 IOS 的备份与恢复、串行线路与静态路由配置、子网掩码与子网划分、静态路由综合配置、PPP 与 RIP 配置、EIGRP 与 OSPF 配置、EIGRP 综合配置、帧中继与子接口配置、OSPF 综合配置、静态路由最佳配置、NAT 及其配置、ACL 基本配置、交换机及其基本配置、VLAN 及其配置、VLAN 间路由与三层交换机配置、VTP 及其配置、STP 及其配置、网络互联综合配置。

教学建议：58 学时或稍多。建议每个项目安排 2 学时，每周学习 2 个项目，计 40 学时；电子资源包括 3 个综合项目，每个项目建议安排 4 学时，每周学习 1 个项目，计 12 学时；总结复习安排 6 学时。

本书既可作为思科网络技术学院 CCNA 1-2 的配套实训教程、高等院校计算机网络及相关专业的教材，也可作为自学者的入门学习资料，还可作为网络工程技术人员的参考书。

未经许可，不得以任何方式复制或抄袭本书之部分或全部内容。
版权所有，侵权必究。

图书在版编目（CIP）数据

路由交换技术项目化教程入门篇 / 袁劲松，胡建荣主编. —北京：电子工业出版社，2020.1
（思科系列丛书）
ISBN 978-7-121-37680-1

Ⅰ. ①路… Ⅱ. ①袁…②胡… Ⅲ. ①计算机网络－路由选择－高等学校－教材②计算机网络－信息交换机－高等学校－教材 Ⅳ. ①TN915.05

中国版本图书馆 CIP 数据核字(2019)第 242920 号

责任编辑：宋　梅
印　　刷：北京天宇星印刷厂
装　　订：北京天宇星印刷厂
出版发行：电子工业出版社
　　　　　北京市海淀区万寿路 173 信箱　邮编：100036
开　　本：787×1092　1/16　印张：16.75　字数：429 千字
版　　次：2020 年 1 月第 1 版
印　　次：2024 年 1 月第 6 次印刷
定　　价：68.00 元

凡所购买电子工业出版社图书有缺损问题，请向购买书店调换。若书店售缺，请与本社发行部联系，联系及邮购电话：（010）88254888，88258888。
质量投诉请发邮件至 zlts@phei.com.cn，盗版侵权举报请发邮件至 dbqq@phei.com.cn。
本书咨询联系方式：mariams@phei.com.cn。

前　　言

本书是计算机网络技术专业建设中课程改革的成果，是在课程团队多年的项目工程设计和教学实施的背景下精心编撰而成的。本书介绍网络互联的基础技术，其内容由精心设计的 20 个实训项目构成，主要包括网络及网络协议的基本知识、路由交换的基本知识、路由器配置项目（14 个）、交换机配置项目（5 个）、网络互联综合项目（1 个）；另附网络构建综合项目（3 个）的电子资源，并通过使用思科网络技术学院网络工具 Packet Tracer 7.2.1 模拟器分步完成；实训项目的安排从简单到复杂、循序渐进、环环相扣，贯穿知识体系，让读者学以致用。

本书在编撰过程中力求突出高职教育的特点，以"理论够用、注重实用"为原则，以培养专业技术能力为重点，将教学内容与职业培养目标相结合，每个案例都可用 Packet Tracer 7.2.1 模拟器模拟，无须昂贵的设备，特别适合作为高职院校相关专业的教材，也非常适合自学。

为了方便专业教学，本书配备了内容丰富的 PPT 课件、习题、教学视频等全套的教辅资源。有需要的读者可以发送邮件到 357775185@qq.com 索取。如果你对网络互联技术感兴趣，那么本书可以为你的成功助力；如果你是一位从事网络互联技术教育工作的专业教师，那么本书及其教辅资源可帮你减少很多繁杂的工作。

本书既可作为思科网络技术学院 CCNA 1-2 的配套实训教程、高等院校计算机网络及相关专业的教材，也可作为自学者的入门学习资料，还可作为网络工程技术人员的参考书。

本书由广东职业技术学院袁劲松、胡建荣组织编写并统稿，参加本书编写和校对工作的还有林冠廷、黎仲恒、杨煜新、杨子帆。感谢思科网络技术学院华南区经理熊露颖、广州市黄埔职业技术学校何力老师、电子工业出版社宋梅编审、思科网络技术学院理事会技术经理沈加航对本书编写工作的大力支持！

本书配套有教学资源 PPT 课件，如有需要，请登录电子工业出版社华信教育资源网（www.hxedu.com.cn），注册后免费下载。

由于时间仓促，作者水平有限，书中难免有不妥和错误之处，恳请读者批评指正。E-mail：357775185@qq.com。

2019 年 12 月
作者

使用说明

为了突出重点、节省篇幅，本书采用了一些特殊格式和简化形式，具体说明如下：
① 添加灰底，表示随后的输入内容需要重点关注。
② 配置过程中只保留操作内容，不显示非必要内容，首次出现的部分内容除外。
③ 有关 PC 或服务器的 IP 地址配置，如下所示。
PC99a 的 IP 地址：10.99.10.254，子网掩码：255.255.255.0，默认网关：10.99.10.1。
④ 有关显示信息，如执行 show ip route 命令显示的内容，只保留需关注的部分。
⑤ 有关网络连通性的测试结果，简化表示，用!!!!!!或!!!!表示"能访问"，用.....表示"不能访问"。
⑥ 设备命名规则：ID 是个人学号的最后两位数字。譬如张三的 ID 是 03，则其路由器以 Rt03a、Rt03b、Rt03c 等命名。书中都以 ID 为 99 作为示例，学生实训时须将示例中所有的 99 都改为自己学号的最后两位数字。

- 路由器以 RtIDa、RtIDb、RtIDc 等命名；
- 二层交换机以 SwIDa、SwIDb、SwIDc 等命名；
- 三层交换机以 MsIDa、MsIDb、MsIDc 等命名；
- 服务器以 SvIDa、SvIDb、SvIDc 等命名；
- 个人计算机以 PCIDa、PCIDb、PCIDc 等命名；
- 笔记本电脑以 LpIDa、LpIDb、LpIDc 等命名；
- 无线设备以 WdIDa、WdIDb、WdIDc 等命名。

⑦ IP 地址分配规则：为了清楚地标明 LAN 的 IP 地址范围，将最高 IP 地址分配给路由器接口、最低 IP 地址分配给 PC（或者相反），剩余的 IP 地址分配给未画出或增加的设备。
⑧ 采用了 C++以//开启注释的风格，用"//注释文字"对有关内容及时地进行简要说明。
⑨ 在本书中，接口的名字均采用简写，其中 Gi 的全称是 GigabitEthernet，Fa 的全称为 FastEthernet，Se 的全称为 Serial，本书后续内容均采用简写来描述网络设备的接口，在程序中，均使用小写。

读者如果需要阅读更详细的内容，可参见本书配套的教辅资源。

目　　录

项目 01　登录路由器 1
　　项目描述 1
　　网络拓扑 1
　　学习目标 1
　　实训任务分解 2
　　知识点介绍 2
　　实训过程 3
　　学习总结 10
　　课后作业 11
　　思考题 11

项目 02　密码恢复和基本配置 12
　　项目描述 12
　　网络拓扑 12
　　学习目标 12
　　实训任务分解 12
　　知识点介绍 13
　　实训过程 13
　　学习总结 18
　　课后作业 19
　　思考题 19

项目 03　配置文件和 IOS 的备份与恢复 20
　　项目描述 20
　　网络拓扑 20
　　学习目标 20
　　实训任务分解 20
　　实训过程 21

学习总结 ·· 28
　　课后作业 ·· 28
　　思考题 ··· 28
项目 04　串行线路与静态路由配置 ·· 29
　　项目描述 ·· 29
　　网络拓扑 ·· 29
　　学习目标 ·· 30
　　实训任务分解 ··· 30
　　实训过程 ·· 30
　　学习总结 ·· 37
　　课后作业 ·· 38
　　思考题 ··· 38
项目 05　子网掩码与子网划分 ·· 39
　　项目描述 ·· 39
　　网络拓扑 ·· 39
　　学习目标 ·· 39
　　实训任务分解 ··· 40
　　实训过程 ·· 40
　　学习总结 ·· 48
　　课后作业 ·· 50
　　思考题 ··· 50
项目 06　静态路由综合配置 ·· 51
　　项目描述 ·· 51
　　网络拓扑 ·· 51
　　学习目标 ·· 51
　　实训任务分解 ··· 52
　　实训过程 ·· 52
　　学习总结 ·· 68
　　课后作业 ·· 68
　　思考题 ··· 69
项目 07　PPP 与 RIP 配置 ··· 70
　　项目描述 ·· 70
　　网络拓扑 ·· 70
　　学习目标 ·· 70
　　实训任务分解 ··· 71
　　知识点介绍 ··· 71
　　实训过程 ·· 71

学习总结···77
　　课后作业···78
　　思考题···78
项目 08　EIGRP 与 OSPF 配置···79
　　项目描述···79
　　网络拓扑···79
　　学习目标···79
　　实训任务分解··80
　　知识点介绍···80
　　实训过程···80
　　学习总结···90
　　课后作业···91
　　思考题···91
项目 09　EIGRP 综合配置···92
　　项目描述···92
　　网络拓扑···92
　　学习目标···92
　　实训任务分解··93
　　知识点介绍···93
　　实训过程···93
　　学习总结···106
　　课后作业···106
　　思考题···107
项目 10　帧中继与子接口配置···108
　　项目描述···108
　　网络拓扑···108
　　学习目标···109
　　实训任务分解··109
　　知识点介绍···109
　　实训过程···109
　　学习总结···121
　　课后作业···121
　　思考题···122
项目 11　OSPF 综合配置···123
　　项目描述···123
　　网络拓扑···123
　　学习目标···124

实训任务分解 ··· 124
　　知识点介绍 ··· 124
　　实训过程 ·· 124
　　学习总结 ·· 135
　　课后作业 ·· 135
　　思考题 ··· 136
项目 12　静态路由最佳配置 ·· 137
　　项目描述 ·· 137
　　网络拓扑 ·· 137
　　学习目标 ·· 137
　　实训任务分解 ··· 138
　　知识点介绍 ··· 138
　　实训过程 ·· 138
　　学习总结 ·· 148
　　思考题 ··· 148
项目 13　NAT 及其配置 ·· 149
　　项目描述 ·· 149
　　网络拓扑 ·· 149
　　学习目标 ·· 150
　　实训任务分解 ··· 150
　　实训过程 ·· 150
　　学习总结 ·· 164
　　课后作业 ·· 165
　　思考题 ··· 165
项目 14　ACL 基本配置 ·· 166
　　项目描述 ·· 166
　　网络拓扑 ·· 166
　　学习目标 ·· 167
　　实训任务分解 ··· 167
　　知识点介绍 ··· 167
　　实训过程 ·· 167
　　学习总结 ·· 181
　　课后作业 ·· 182
　　思考题 ··· 183
项目 15　交换机及其基本配置 ·· 184
　　项目描述 ·· 184
　　网络拓扑 ·· 184

学习目标	184
实训任务分解	185
知识点介绍	185
实训过程	188
学习总结	195
课后作业	195
思考题	195

项目 16　VLAN 及其配置 …… 196

项目描述	196
网络拓扑	196
学习目标	196
实训任务分解	197
知识点介绍	197
实训过程	202
学习总结	205
课后作业	205
思考题	206

项目 17　VLAN 间路由与三层交换机配置 …… 207

项目描述	207
学习目标	208
实训任务分解	208
知识点介绍	208
实训过程	208
学习总结	215
课后作业	215
思考题	215

项目 18　VTP 及其配置 …… 216

项目描述	216
网络拓扑	216
学习目标	216
实训任务分解	217
知识点介绍	217
实训过程	219
学习总结	233
课后作业	233
思考题	233

项目 19 STP 及其配置	234
项目描述	234
网络拓扑	234
学习目标	234
实训任务分解	235
知识点介绍	235
实训过程	239
学习总结	246
课后作业	246
思考题	246
项目 20 网络互联综合配置	247
项目描述	247
网络拓扑	247
学习目标	247
实训任务分解	248
具体要求	248
实训过程	249
学习总结	254
课后作业	255
思考题	255

项目 01　登录路由器

项目描述

若你是某单位的网络管理员，你需要查看路由器的各种信息，或者需要修改路由器的现有配置，或者需要对新购买的路由器进行配置，由于路由器没有键盘和显示器，因此，你首先要借助计算机登录到路由器上。

网络拓扑

网络拓扑如图 1-1 所示。

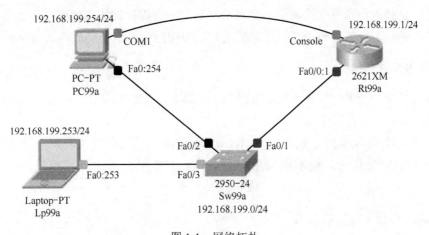

图 1-1　网络拓扑

学习目标

- 能从 Console 端口登录路由器；
- 能用 Telnet 远程登录路由器；
- 初步了解路由器的命令行界面（CLI）及其应用；
- 初步掌握 Cisco Packet Tracer 模拟器的使用方法。

实训任务分解

① 创建网络拓扑连接。
② 通过 Console 端口，实现初次本地登录。
③ 通过网络连接，实现 Telnet 远程登录。
④ 完成配置后，实现再次本地登录。
⑤ 保存配置和网络拓扑。

知识点介绍

1．为什么要登录？如何登录？

在构建网络时，首先要根据需求做好规划、选购设备，再用适当的线缆把计算机、路由器、交换机等网络设备连接起来，然后还要对它们进行适当的配置和调试，直到满足用户的通信需求。

由于路由器和交换机都没有键盘和显示器，因此，要想对它们进行配置，就必须先借助计算机登录到路由器或交换机上。登录成功后，方可借助计算机的键盘和显示器对路由器或交换机进行配置或管理。必须登录也是对网络中路由器或交换机的一种保护。

2．登录配置环境

登录方式可分为本地登录和远程登录两类。本地登录就是从 Console 端口登录路由器或交换机，其他方式都属于远程登录方式。

我们主要学习和掌握本地登录和 Telnet 远程登录。远程登录需要先连通网络，因此，进行初始化配置时只能本地登录。路由器和交换机的登录方式相似，下面以路由器为例。

3．通过 Console 端口登录

第一步是用反转线把路由器的 Console 端口与计算机的串口相连接。
第二步是打开电源开关，加电启动计算机和路由器。
第三步是启动超级终端登录到路由器。其过程如下：在 Windows 中的"开始"→"程序"→"附件"→"通信"菜单下打开"超级终端"程序，出现图 1-2 窗口。在"名称"对话框中输入名称，例如"Router"，单击【确定】按钮。

当出现图 1-3 所示窗口时，在"连接时使用"下拉菜单中选择计算机的 COM1 口，单击【确定】按钮，完成超级终端端口设置。按图 1-4 设置超级终端端口属性，然后单击【确定】按钮。按【回车】键，查看超级终端窗口中是否出现路由器提示符或其他字符，如果

出现提示符或者其他字符，则说明已经连接上路由器，可以登录了。

设备间用反转线连接，所谓反转（Rollover）线就是线两端的 RJ-45 接头上的线序是相反的，反转线的线序图如图 1-5 所示。

图 1-2　超级终端

图 1-3　超级终端端口设置

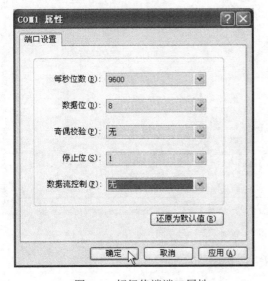

图 1-4　超级终端端口属性

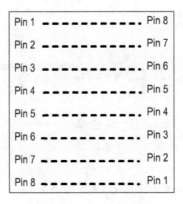

图 1-5　反转线的线序图

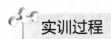
实训过程

完成下面的模拟实训后，教师应在真实的环境下再演示一遍，以便让学生更好地了解模拟与真实的区别和相同之处，更好地理解和掌握所学的内容。

1. 创建网络拓扑连接

由于条件有限，很多实训需要借助于模拟仿真软件。Cisco 公司开发了一套非常逼真好用的模拟仿真软件，它就是 Packet Tracer。我们借助于它可以随时随地完成网络实训，只要有计算机就行。它使我们不再依赖于因代价昂贵而资源有限的实训室。这是真正的好东西，相信大家都会喜欢的。

Packet Tracer 是由 Cisco 公司发布的一个辅助学习 CCNA 课程的模拟软件。我们可用它建立网络拓扑，配置其中的设备，测试网络通信。它还可图示数据包在网络中传送的详细处理过程，让我们观察到网络的实时运行情况。Packet Tracer 为我们提供了方便的实训环境和最接近真实环境的体验和感受。Packet Tracer 7.2.1 的界面如图 1-6 所示。

图 1-6 Packet Tracer 7.2.1 的界面

请在模拟器中创建如图 1-1 所示的网络拓扑。

2．通过 Console 端口，实现初次本地登录

新路由器只能通过 Console 端口实现本地登录，登录时无须输入任何口令。登录已配置过的路由器一般都需要输入相应的口令。下面是模拟本地登录的过程。

（1）打开 Desktop 选项卡

在模拟器中双击 PC99a，在弹出的操作界面单击 Desktop 选项卡，显示界面如图 1-7 所示。

（2）单击 Terminal 图标

单击 Terminal 图标，显示界面如图 1-8 所示。

项目 01　登录路由器

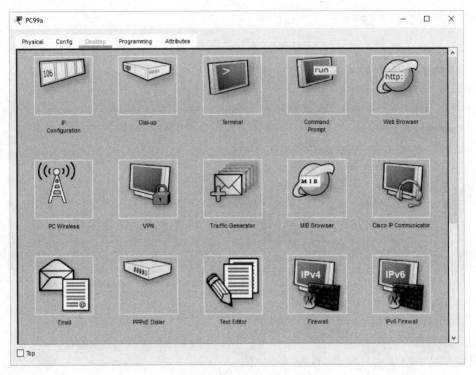

图 1-7　Desktop 选项卡

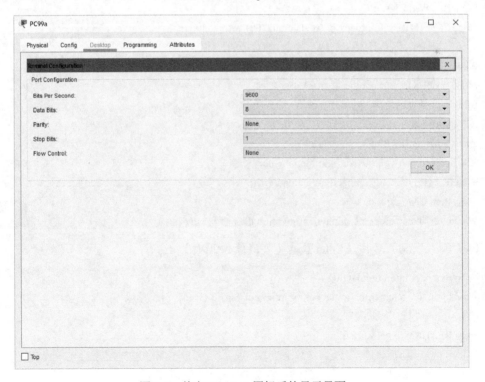

图 1-8　单击 Terminal 图标后的显示界面

5

(3)单击【OK】按钮

单击【OK】按钮,将显示如下的信息:

Readonly ROMMON initialized
Self decompressing the image :
[OK]
Restricted Rights Legend
Use, duplication, or disclosure by the Government is
subject to restrictions as set forth in subparagraph
(c) of the Commercial Computer Software - Restricted
Rights clause at FAR sec. 52.227-19 and subparagraph
(c) (1) (ii) of the Rights in Technical Data and Computer
Software clause at DFARS sec. 252.227-7013.
cisco Systems, Inc.
170 West Tasman Drive
San Jose, California 95134-1706

Cisco Internetwork Operating System Software
IOS (tm) C2600 Software (C2600-I-M), Version 12.2(28), RELEASE SOFTWARE (fc5)
Technical Support: http://www.cisco.com/techsupport
Copyright (c) 1986-2005 by cisco Systems, Inc.
Compiled Wed 27-Apr-04 19:01 by miwang
Cisco 2621 (MPC860) processor (revision 0x200) with 253952K/8192K bytes of memory
.
Processor board ID JAD05190MTZ (4292891495)
M860 processor: part number 0, mask 49
Bridging software.
X.25 software, Version 3.0.0.
2 FastEthernet/IEEE 802.3 interface(s)
32K bytes of non-volatile configuration memory.
63488K bytes of ATA CompactFlash (Read/Write)
--- System Configuration Dialog ---
Would you like to enter the initial configuration dialog? [yes/no]:

(4)输入"no"后按【Enter】键(可连续按几次)

--- System Configuration Dialog ---
Would you like to enter the initial configuration dialog? [yes/no]:   no
//输入 no,不进入初始配置对话
Press RETURN to get started!
Router>

当最后一行的命令提示符出现时,就说明已成功登录了。

项目 01　登录路由器

3．通过网络连接，实现 Telnet 远程登录

远程登录以网络通信为前提，因此，需要在上面成功登录的基础上继续做一些准备工作，以连通计算机与路由器。

（1）在 PC99a 上配置路由器

```
Router> en
Router# conf  t
Enter configuration commands, one per line. End with CNTL/Z.
Router(config)#
Router(config)# hostname Rt99a              //将该路由器命名为 Rt99a，其中 99 为学号最后两位数字
Rt99a(config)# enable password 99secret     //设置 enable 口令为 99secret
Rt99a(config)# line vty 0 4
Rt99a(config-line)# password 99vty04        //配置远程登录口令，以开启远程登录功能
Rt99a(config-line)# login
Rt99a(config-line)# exit
Rt99a(config)#
Rt99a(config)# int fa0/0
Rt99a(config-if)# ip address 192.168.199.1 255.255.255.0    //配置网络端口
Rt99a(config-if)# no shutdown                               //开启端口
Rt99a(config-if)#
%LINK-5-CHANGED: Interface FastEthernet0/0, changed state to up
%LINEPROTO-5-UPDOWN: Line protocol on Interface FastEthernet0/0, changed state to up
Rt99a(config-if)# end                                       //返回特权模式
```

（2）配置计算机的 IP 地址

如图 1-9 所示，以 Lp99a 为例配置计算机的 IP 地址。

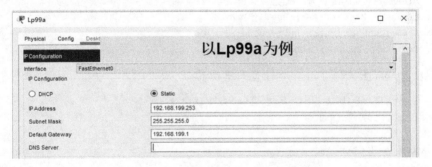

图 1-9　以 Lp99a 为例配置计算机的 IP 地址

① Lp99a 的配置。
- IP 地址：192.168.199.253；
- 子网掩码：255.255.255.0；
- 默认网关：192.168.199.1。

② PC99a 的配置。
- IP 地址：192.168.199.254；

7

- 子网掩码：255.255.255.0；
- 默认网关：192.168.199.1。

完成路由器和计算机的配置后，模拟远程登录拓扑如图 1-10 所示，该图显示了网络状态。

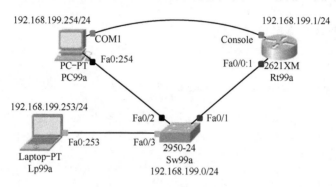

图 1-10　模拟远程登录拓扑

（3）在 Lp99a 上测试与路由器的连通性

```
C:\> ping 192.168.199.1                               //可通

Pinging 192.168.199.1 with 32 bytes of data:

Reply from 192.168.199.1: bytes=32 time<1ms TTL=255
Reply from 192.168.199.1: bytes=32 time=3ms TTL=255
Reply from 192.168.199.1: bytes=32 time<1ms TTL=255
Reply from 192.168.199.1: bytes=32 time<1ms TTL=255

Ping statistics for 192.168.199.1:
Packets: Sent = 4, Received = 4, Lost = 0 (0% loss),
Approximate round trip times in milli-seconds:
Minimum = 0ms, Maximum = 3ms, Average = 0ms
```

（4）在 Lp99a 上采用 Telnet 远程登录路由器

```
C:\>  telnet 192.168.199.1      //远程登录 Rt99a 的命令
Trying 192.168.199.1 ...Open

User Access Verification

Password:                       //远程登录时必须输入口令 99vty04
Rt99a>
Rt99a>  en
Password:                       //进入特权模式时必须输入口令 99secret
Rt99a#
```

以上操作说明远程登录成功了。

4．完成配置后，实现再次本地登录

在 PC99a 上进行以下操作。

（1）保存配置、退出特权模式、退出登录状态

保存配置、退出特权模式、退出登录状态，然后重新登录路由器。

```
Rt99a#   copy   run   start              //保存配置
Destination filename [startup-config]?
Building configuration...
[OK]
Rt99a#   disable                         //退出特权模式
Rt99a>
Rt99a>   logout                          //退出登录状态
```

（2）再次本地登录

```
Press RETURN to get started.
User Access Verification
Password:                                //必须输入口令 99con 方可登录
Rt99a>
Rt99a>   enable                          //进入特权模式
Password:                                //必须输入口令 99secret 方可进入特权模式
Rt99a#
```

当再次登录时，必须输入正确的口令才能登录。

5．保存配置和网络拓扑

在每次实训的最后，都要保存配置和网络拓扑。以本次为例，具体操作如下：

① 在做了适当配置的各路由器和交换机上执行相关命令保存配置，例如：

Rt99a#   copy   run   start   或   Rt99a#   write

② 在文件菜单中选择退出时，将弹出是否保存修改的对话框，选择 Yes，将网络拓扑和配置保存在文件名为 P01001.pkt 的文件中，以备在下一个项目中使用，如图 1-11 所示。

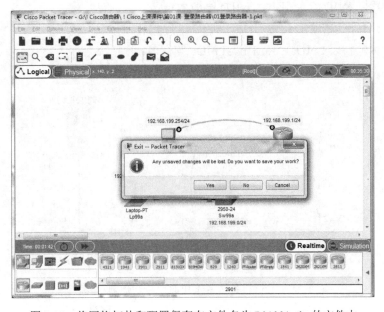

图 1-11　将网络拓扑和配置保存在文件名为 P01001.pkt 的文件中

学习总结

在本项目中我们学会了如何完成本地登录，如何开启远程登录功能，以及如何完成远程登录。我们还初步学会了模拟器的使用方法。通过本项目的学习和实训，我们知道，当路由器设置了口令以后，要想登录路由器就必须输入正确的口令。这为网络中的路由器提供了必要的保护，登录及其口令的作用也正在于此。

当初始化路由器时，只能从 Console 端口登录路由器。在路由器允许远程登录并能连通路由器的情况下，方可远程登录路由器。路由器和交换机是构建网络时最常用的设备，通常用交换机构建一个以太网，用路由器连接不同的网络。交换机的全部接口通常属于一个网络，路由器的每个接口都必须分属于不同的网络。

同一以太网上的所有主机的 IP 地址的网络号必须相同。在命令行提示符下输入问号会显示当前可用的命令，要了解命令的下一个参数也可输入一个问号，按"Tab"键可以补齐单词剩下的字符。

1. 设备命名规则

ID 是各自学号的最后两位数字。例如，张三的 ID 是 03，则其路由器以 Rt03a、Rt03b、Rt03c 等命名。图 1-12 以 ID 为 99 为示例，给出了设备命名和 IP 地址分配规则示例。

① 路由器以 RtIDa、RtIDb、RtIDc 等命名；
② 二层交换机以 SwIDa、SwIDb、SwIDc 等命名；
③ 三层交换机以 MsIDa、MsIDb、MsIDc 等命名；
④ 服务器以 SvIDa、SvIDb、SvIDc 等命名；
⑤ 个人计算机以 PCIDa、PCIDb、PCIDc 等命名；
⑥ 笔记本电脑以 LpIDa、LpIDb、LpIDc 等命名；
⑦ 无线设备以 WdIDa、WdIDb、WdIDc 等命名。

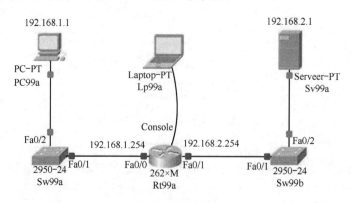

图 1-12 设备命名和 IP 地址分配规则示例

项目 01　登录路由器

2．IP 地址分配规则

为了清楚地标明 LAN 的 IP 地址范围，将最高 IP 地址分配给路由器接口，将最低 IP 地址分配给 PC（或者相反），剩余的 IP 地址分配给未画出或将来的设备。在本书中，以后都要遵循这样的设备命名和 IP 地址分配规范。

图 1-12 中的 IP 地址分配表如表 1-1 所示。

表 1-1　图 1-12 中的 IP 地址分配表

	最低 IP 地址	最高 IP 地址	广播地址
网络 192.168.1.0	192.168.1.1	192.168.1.254	192.168.1.255
IP 地址对应的设备	PC99a	Rt99a 的 Fa0/0	
网络 192.168.2.0	192.168.2.1	192.168.2.254	192.1682.2.255
IP 地址对应的设备	Sv99a	Rt99a 的 Fa0/1	

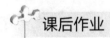

课后作业

完成本项目中模拟仿真实训，并将实训过程的截图按顺序粘贴到一个 Word 文件里，再加上适当的文字用以说明你对它的理解、认识、领悟等，作为实训报告上交，实训报告的封面另发。

ID 是个人学号的最后两位数字，实训报告一律以"ID 姓名项目号.doc"命名，网络拓扑及其配置也以"ID 姓名项目号.pkt"为文件名保存并上交。例如，张三的 ID 为 03，则他的文件名为"03 张三 01.doc"和"03 张三 01.pkt"。

为了杜绝抄袭作业行为，本书为网络设备的命名制定了统一的规范；为了清楚地标明 IP 地址的取值范围，本书也为 IP 地址的分配制定了统一的规范。以后的作业都要遵循上述规范；而且对任何设备进行配置的第一项内容都是配置主机名即设备名。学生实训时必须将 99 改为自己的学号的最后两位数字。

思考题

为了查询或修改路由器的配置，要想登录路由器时却发现口令忘记了，那我们又该怎么办呢？

在下一个项目中我们将学习解决办法。

项目 02　密码恢复和基本配置

项目描述

若你是某单位的网络管理员，因不知某路由器的口令而不能登录，那你就必须恢复口令；为了便于管理，你还需要对路由器做一些基本的配置工作，如配置路由器的主机名并完成同步信息；取消超时退出和域名解析；开启并控制远程登录功能等。

网络拓扑

网络拓扑如图 2-1 所示。

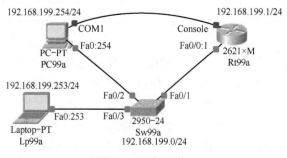

图 2-1　网络拓扑

学习目标

- 掌握恢复路由器密码的方法。
- 掌握配置路由器基本信息的方法。
- 掌握查看路由器基本信息的方法。
- 初步掌握路由器的命令行界面（CLI）。
- 逐步熟悉 Cisco Packet Tracer 模拟器的使用方法。

实训任务分解

① 恢复路由器密码。

② 完成路由器基本配置。
③ 查看路由器基本信息。

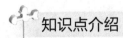

知识点介绍

由于各种原因，网络管理员经常会遇到忘记密码（口令）或不知密码的情况。那怎么办呢？只要能接触路由器的电源开关，就有办法恢复或重设口令。

网络管理员为了便于管理，一般也需要修改路由器的一些出厂配置。

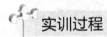

实训过程

1. 恢复路由器密码

当因不知路由器的口令或密码而不能登录时，就必须恢复口令或密码。假设我们忘记了项目 01 中配置的路由器的口令，则恢复路由器口令的步骤和方法如下。

（1）忘记登录口令面对的困境

把项目 01 保存的 P01001.pkt 文件先另存为文件名为 P02001.pkt 的文件。张三的文件应该先另存为文件名为 03 张三 02.pkt 的文件。双击文件 P02001.pkt 并在 PC99a 上启动超级终端。

因忘记登录口令而不能登录系统的困境如图 2-2 所示。

图 2-2　忘记登录口令

（2）进入 rommon 模式

按图 2-3 所示，关闭路由器电源后重新开机，在控制台显示启动过程的 60 秒内，赶快按【Ctrl+Break】键(或【Ctrl+C】键)中断路由器的启动过程，进入 rommon 模式，如图 2-4 所示，执行如下命令：

```
rommon 1>                              //输入以下命令改变配置寄存器的值为 0x2142
rommon 1>   confreg  0x2142            //使路由器重启时绕过口令检测
rommon 2>   reset                      //输入重启命令
```

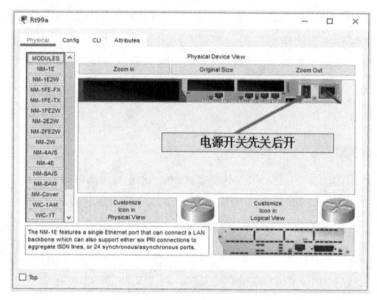

图 2-3　关闭路由器电源后重新开机

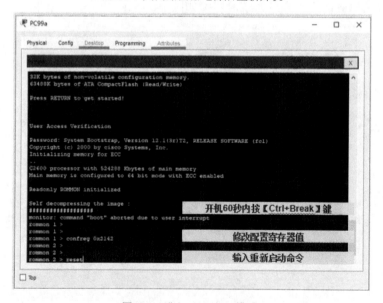

图 2-4　进入 rommon 模式

（3）进入 setup 配置模式

按【Enter】键，路由器重启后会直接进入 setup 配置模式，输入"no"后按【Enter】

键退出 setup 模式,显示提示符。重新开机,如图 2-5 所示。

图 2-5 重新开机

(4)恢复原有配置

这一步对保留其他配置很关键。输入以下命令:

Router> enable
Router# copy startup-config running-config
Destination filename [running-config]?

按【Enter】键,将配置文件从 NVRAM 复制到 RAM 中,如图 2-6 所示为保存配置信息。在此基础上再重设新密码,就能保留原有的其他配置。

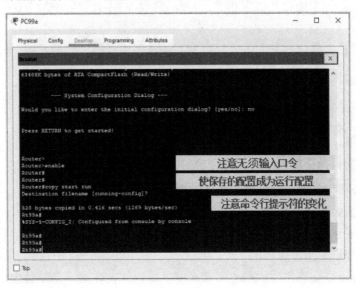

图 2-6 保存配置信息

（5）设置新口令

① 进入全局配置模式。

Rt99a# configure terminal

② 将特权模式口令设置为新口令 99cisco。

Rt99a(config)# enable secret 99cisco

③ 将本地登录口令设置为新口令 99con0。

Rt99a(config)# line con 0
Rt99a(config-line)# password 99con0
Rt99a(config-line)# login

（6）恢复配置寄存器正常值（0x2102）

Rt99a(config-line)# exit
Rt99a(config)# config-register 0x2102
Rt99a(config)#

（7）保存配置

Rt99a# copy running-config startup-config
Destination filename [startup-config]?

按【Enter】键，将配置文件从 RAM 复制到 NVRAM 中。若没有（4）中的 copy 命令，则执行上面的命令后，原有的其他配置都将被改回出厂配置。

（8）重启路由器

Rt99a# reload

按【Enter】键后路由器将重启。重启后，请检查路由器的新口令和其他配置是否正常。若正常，则路由器的本地登录口令和进入特权模式的口令就已完全恢复，一切都已尽在你的掌控之中了。

注意：不能接触路由器开关的人，以上办法是不可行的。

（9）保存拓扑

在文件菜单中选择退出时，会弹出是否保存修改的对话框，选择 Yes，将网络拓扑和配置保存为文件名为 P02001.pkt 的文件，以备在下一个项目中使用。

2. 完成路由器基本配置

在上面的操作中可能你已发现有以下问题：

① 如果一段时间没有操作，则会话连接会自动断开。

② 会不断显示信息，干扰命令输入。

③ 登录口令都是明码保存的。

这些问题可通过一些常用的基本配置来解决。例如，在上一个项目中，可通过以下配置解决输错的命令被当作域名解析的问题：

Rt99a(config)# no ip domain-lookup

用一台新的路由器来完成基本配置练习,基本配置练习网络拓扑如图 2-7 所示。由于只是练习,因此不必保存配置。实际完成的基本配置如图 2-8 所示。

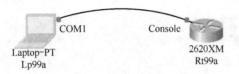

图 2-7　基本配置练习网络拓扑

图 2-8　实际完成的基本配置

所谓基本配置的界限比较模糊,主要包括以下内容:

(1) 配置路由器主机名

Router(config)#　hostname　Rt99a

(2) 配置本地登录口令

Rt99a(config)#　line con　0
Rt99a(config-line)#　password　99con0
Rt99a(config-line)#　login

(3) 配置远程登录口令和开启远程登录

Rt99a(config)#　line　vty　0　4

```
Rt99a(config-line)#   password   99vty04
Rt99a(config-line)#   login
```

（4）配置进入特权模式的口令

```
Rt99a(config)#   enable   password   99pass      //口令明文保存
Rt99a(config)#   enable   secret   99secret      //口令密文保存
```

（5）加密保存所有口令

```
Rt99a(config)#   service   password-encryption   //使所有口令都加密保存
```

（6）同步提示信息，避免其干扰命令的输入

```
Rt99a(config-line)#   logging   synchronous
```

（7）取消超时自动退出（logout）

```
Rt99a(config-line)#   no   exec-timeout
```
或
```
Rt99a(config-line)# exec-timeout   0   0
```

（8）取消域名解析

```
Rt99a(config)#   no   ip   domain-lookup
```

（9）设置路由器的日期和时间

```
Rt99a#   clock   set   10:23:45   3   Sep   2013
```

3．查看路由器基本信息

进入特权模式后，可用 show 命令查看各种信息：

```
# show   running-config          //查看运行配置
# show   startup-config          //查看启动配置
# show   interfaces              //查看接口信息
# show   version                 //查看 IOS 版本信息
# show   flash                   //查看闪存内容
# show   protocols               //查看网络协议
# show   processes               //查看进程
# show   ip   arp                //查看 ARP 信息
```

学习总结

在能够物理接触路由器的情况下，即使忘记了密码，我们仍可按上述的办法重新控制路由器。给路由器取个描述性的名字，有利于我们更好地了解和管理网络。管理好口令，可以有效保护网络中的路由器。使提示信息同步输出、取消超时退出、取消域名解析等，可以使我们输入命令时感觉更好一些。

路由器的基本信息对我们了解路由器是非常重要的，据此我们可以判断 IOS 版本是否是最新的、有哪些网络接口可用或者有哪些网络接口出现了问题。可以通过适当的 show 命令来了解路由器。请查询路由器查找 IOS 的流程和 show 命令能显示哪些信息。

路由器密码恢复步骤如下所述：

① 启动路由器，60 秒内按下【Ctrl+Break】键。
② rommon> confreg 0x2142。
③ rommon> reset。
④ router# copy startup-config running-config。
⑤ router(config)# enable secret newpassword1。
⑥ router(config)# line con 0。
　router(config-line)# password newpassword2。
　router(config-line)# login。
⑦ router(config)# config-register 0x2102。
⑧ router# copy running-config startup-config。
⑨ router# reload。

上面的第④步非常重要，缺了这步会丢失以前的所有配置。

课后作业

完成路由器密码恢复的模拟实训，将实训过程的截图按顺序粘贴到一个 Word 文件里，再加上适当的文字用以说明你对它的理解、认识、领悟等；总结本次实训所需要的主要命令及其作用，作为实训报告上交；实训内容如前面的截图所示。

根据项目 01 中的路由器命名规则，务必把路由器命名为 RtIDa，其中 ID 为各自学号的最后两位数字。实训报告一律以"ID 姓名项目号.doc"命名，网络拓扑及其配置也以"ID 姓名项目号.pkt"为文件名保存并上交。例如，张三的 ID 为 03，他的文件名为"03 张三 02.doc"和"03 张三 02.pkt"。

思考题

要想备份路由器的配置文件和 IOS，怎么做？要想恢复被误删的路由器配置文件或需要升级 IOS，怎么做？

在下一个项目中我们将学习解决办法。

项目 03　配置文件和 IOS 的备份与恢复

项目描述

若你是某单位的网络管理员，为了安全，你需要备份路由器的配置文件和 IOS；当路由器的配置文件和 IOS 损坏时，你需要恢复路由器的配置文件和 IOS。

网络拓扑

网络拓扑如图 3-1 所示。

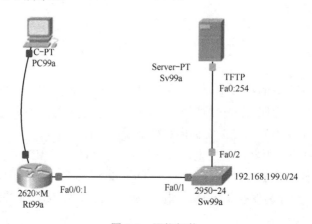

图 3-1　网络拓扑

学习目标

- 掌握配置文件和 IOS 的备份方法。
- 掌握恢复配置文件和 IOS 的方法。
- 能安装和使用 TFTP 服务器。

实训任务分解

① 备份配置文件和 IOS。

项目 03　配置文件和 IOS 的备份与恢复

② 恢复配置文件和 IOS。

实训过程

1．备份配置文件和 IOS

从安全的角度考虑，为了预防丢失或损坏，任何有用的文件都应做好备份。

计算机网络配置、调试好以后，所有路由器的配置文件都应做备份。IOS 虽然可以从 Cisco 公司官方网站下载，但每个 IOS 版本最好也做一个备份。

备份配置文件和 IOS 的命令如下：

```
Rt99a#  copy  running-config  tftp          //备份运行配置文件的命令
Rt99a#  copy  flash  tftp                    //备份 IOS 的命令
```

有了这些备份，可尽快恢复损坏的文件。

（1）完成路由器 Rt99a 基本配置

```
Router(config)# host Rt99a
Rt99a(config)# no ip domain-lookup
Rt99a(config)# enable secret 99secret
Rt99a(config)# service password-encryp
Rt99a(config)# line con 0
Rt99a(config-line)# logg sync
Rt99a(config-line)# exec-timeout 0 0
Rt99a(config-line)# line vty 0 4
Rt99a(config-line)# password 99vty04
Rt99a(config-line)# login
Rt99a(config-line)# exec-timeout 0 0
Rt99a(config-line)# exit
```

（2）配置网络接口并保存配置

```
Rt99a(config)# int fa0/0
Rt99a(config-if)# ip addr 192.168.199.1 255.255.255.0
Rt99a(config-if)# no shutdown              //以上 3 行命令：配置接口
Rt99a(config-if)# end
Rt99a# copy run start                       //保存配置
```

（3）保存文件

单击【保存】按钮，将相关配置保存到文件 P03001.pkt 中以备用。

关于"备用"的说明：P03001.pkt 是本项目实训过程中生成的第一个文件，为了便于学习，将其保存以备当下一个步骤出错时可以从这个文件开始，而不至于从头开始。P03001 中的前三位是项目编号，表示项目 03，后三位是保存文件编号。

（4）安装、配置 TFTP 服务器

在真实环境中，下载 TFTP_Server_TFTPDWIN_v0.4.2.rar，解压后，双击 tftpdwin.exe 图标，按向导安装、配置，保证它处于运行状态即可，非常简单。

在模拟器中，TFTP 服务器已经配置和启动，且已保存了很多版本的 IOS 文件。

（5）查看 TFTP 服务器基本信息

在 Sv99a 上查看 TFTP 服务器基本信息，如图 3-2 所示。

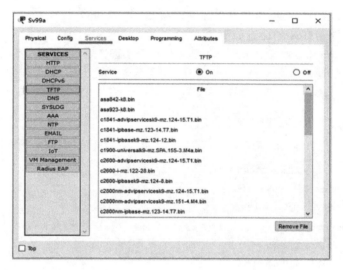

图 3-2　在 Sv99a 上查看 TFTP 服务器基本信息

（6）配置 Sv99a

- IP 地址：192.168.199.254；
- 子网掩码：255.255.255.0；
- 默认网关：192.168.199.1。

（7）在 Rt99a 上测试网络连通性

```
Rt99a# ping 192.168.199.254
.!!!!
```

上面的测试结果证明：Rt99a 已与 TFTP 服务器 Sv99a 连通。

（8）备份配置文件

```
Rt99a# copy run tftp                        //备份运行配置文件的命令
Address or name of remote host []? 192.168.199.254
Destination filename [Rt99a-confg]? Rt99a-config
Writing running-config...!!
[OK - 587 bytes]
587 bytes copied in 0 secs
```

（9）确认已备份配置文件

在 Sv99a 上确认已备份配置文件，如图 3-3 所示。

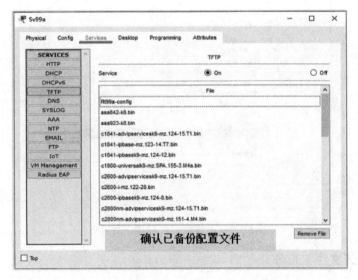

图 3-3　在 Sv99a 上确认已备份配置文件

（10）查看和备份 IOS

```
Rt99a# show flash
System flash directory:
File Length Name/status
3 5571584 c2600-i-mz.122-28.bin                    //IOS
2 28282 sigdef-category.xml
1 227537 sigdef-default.xml
[5827403 bytes used, 58188981 available, 64016384 total]
63488K bytes of processor board System flash (Read/Write)

Rt99a# copy flash tftp                             //备份 IOS 的命令
Source filename []? c2600-i-mz.122-28.bin
Address or name of remote host []? 192.168.199.254
Destination filename [c2600-i-mz.122-28.bin]? c2600-i-mz.122-28.bin.bak
Writing c2600-i-mz.122-28.bin...!!!!!!!!!!!!!!!!!!!!!!!!!!!!!!!!!!!!!!!!!!!!!!!!!!!!!!!!!!!!!!!!!!!!!!!!!!!!!!!!!!!!!!!!!!!!!!!!!!!!!!!!!!!!!!!!!!!!!
[OK - 5571584 bytes]
5571584 bytes copied in 0.206 secs (6197168 bytes/sec)
```

（11）确认已备份 IOS

在 Sv99a 上确认已备份 IOS，如图 3-4 所示。

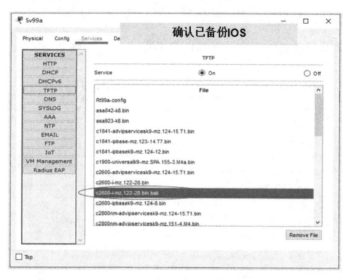

图 3-4　在 Sv99a 上确认已备份 IOS

2．恢复配置文件和 IOS

实训环境与备份时相同。由于配置文件丢失，所以要先使路由器与 TFTP 服务器的网络恢复连通，然后才能恢复配置文件和 IOS。恢复运行配置文件和 IOS 的命令如下：

```
Router# copy   tftp   run           //恢复运行配置文件的命令
Rt99a# copy   tftp   flash          //恢复 IOS 的命令
```

（1）恢复配置文件

先使配置文件丢失或损坏，再使配置文件恢复正常。

① 在 Rt99a 上故意删除配置文件后重启（模拟配置文件丢失或损坏）。

```
Rt99a# erase startup-config         //删除配置文件的命令
Erasing the nvram filesystem will remove all configuration files! Continue? [confirm]
[OK]                                //按【Enter】键确认便开始删除
Erase of nvram: complete
%SYS-7-NV_BLOCK_INIT: Initialized the geometry of nvram
Rt99a# reload                       //重启命令
Proceed with reload? [confirm]      //按【Enter】键便重新启动
```

② 在 Rt99a 上确认配置文件丢失。

```
Copyright (c) 1986-2005 by cisco Systems, Inc.
Compiled Wed 27-Apr-04 19:01 by miwang
Cisco 2620 (MPC860) processor (revision 0x200) with 253952K/8192K bytes of memory
.
Processor board ID JAD05190MTZ (4292891495)
M860 processor: part number 0, mask 49
Bridging software.
X.25 software, Version 3.0.0.
```

```
1 FastEthernet/IEEE 802.3 interface(s)
32K bytes of non-volatile configuration memory.
63488K bytes of ATA CompactFlash (Read/Write)

//若有启动配置文件,则上面能看到该文件且不会出现下面的系统配置对话。
--- System Configuration Dialog
Would you like to enter the initial configuration dialog? [yes/no]: n
Press RETURN to get started!
Router>
```

③ 将网络恢复到连通状态。

```
Router> en
Router# conf t
Enter configuration commands, one per line. End with CNTL/Z.
Router(config)# int fa0/0
Router(config-if)# ip addr 192.168.199.1 255.255.255.0
Router(config-if)# no shut
Router(config-if)# end
Router# ping 192.168.199.254          //测试服务器的连通性
Type escape sequence to abort.
Sending 5, 100-byte ICMP Echos to 192.168.199.254, timeout is 2 seconds:
.!!!!
Success rate is 80 percent (4/5), round-trip min/avg/max = 0/1/4 ms
```

④ 在 Rt99a 上恢复配置文件。

```
Router# copy tftp run                                //恢复运行配置文件的命令
Address or name of remote host []? 192.168.199.254
Source filename []? Rt99a-config
Destination filename [running-config]?               //按【Enter】键便开始恢复
Accessing tftp://192.168.199.254/Rt99a-config...
Loading Rt99a-config from 192.168.199.254: !
[OK - 587 bytes]
587 bytes copied in 0.001 secs (587000 bytes/sec)
Rt99a#                                               //注意提示符的变化
%SYS-5-CONFIG_I: Configured from console by console
Rt99a# copy run start                                //保存运行配置文件的命令
Destination filename [startup-config]?               //按【Enter】键便开始保存
Building configuration...
[OK]                                                 //保存完毕
```

（2）恢复配置 IOS

恢复 IOS 分以下两种情况:
① 误删除 IOS 后还没重启。这种情况的恢复就比较容易。
② IOS 损坏后已重启。这种情况的恢复则比较麻烦。

警告：不要在工作路由器上故意删除 IOS！

① 在 Rt99a 上查看并故意删除 IOS。

```
Rt99a# show flash                                           //查看 IOS
System flash directory:
File Length Name/status
3 5571584 c2600-i-mz.122-28.bin                             //IOS
2 28282 sigdef-category.xml
1 227537 sigdef-default.xml
[5827403 bytes used, 58188981 available, 64016384 total]
63488K bytes of processor board System flash (Read/Write)

Rt99a# delete flash:c2600-i-mz.122-28.bin                   //删除 IOS 的命令
Delete filename [c2600-i-mz.122-28.bin]?                    //按【Enter】键确认
Delete flash:/c2600-i-mz.122-28.bin? [confirm]              //按【Enter】键再次确认便开始删除
```

注意：删除后不要重启！

② 在 Rt99a 上确认已删除 IOS 并恢复。

```
Rt99a# show flash                                           //查看
System flash directory:
File Length Name/status
2 28282 sigdef-category.xml
1 227537 sigdef-default.xml
[255819 bytes used, 63760565 available, 64016384 total]
63488K bytes of processor board System flash (Read/Write)
//以上输出中未见 IOS
Rt99a# copy tftp flash                                      //恢复 IOS 的命令
Address or name of remote host []? 192.168.199.254
Source filename []? c2600-i-mz.122-28.bin.bak
Destination filename [c2600-i-mz.122-28.bin.bak]? c2600-i-mz.122-28.bin
Accessing tftp://192.168.199.254/c2600-i-mz.122-28.bin.bak...
Loading c2600-i-mz.122-28.bin.bak from 192.168.199.254: !!!!!!!!!!!!!!!!!!!!!!!!!!!!!!!!!!!!!!!!!!!!!!!!!!
!!!!!!!!!!!!!!!!!!!!!!!!!!!!!!!!!!!!!!!!!!!!!!
[OK - 5571584 bytes]                                        //恢复完毕
5571584 bytes copied in 0.212 secs (6021776 bytes/sec)
```

③ 在 Rt99a 上确认 IOS 已恢复。

```
Rt99a# show flash                                           //查看
System flash directory:
File Length Name/status
4 5571584 c2600-i-mz.122-28.bin                             //IOS
2 28282 sigdef-category.xml
1 227537 sigdef-default.xml
[5827403 bytes used, 58188981 available, 64016384 total]
63488K bytes of processor board System flash (Read/Write)
Rt99a#
```

④ 在 Rt99a 上故意删除 IOS 并重启。
注意：千万不要在工作路由器上做！

Rt99a# delete flash:c2600-i-mz.122-28.bin	//删除 IOS 的命令
Delete filename [c2600-i-mz.122-28.bin]?	//按【Enter】键确认
Delete flash:/c2600-i-mz.122-28.bin? [confirm]	//按【Enter】键再次确认便开始删除
Rt99a# show flash	//查看发现 IOS 已被删除
System flash directory:	
File Length Name/status	
2 28282 sigdef-category.xml	
1 227537 sigdef-default.xml	
[255819 bytes used, 63760565 available, 64016384 total]	
63488K bytes of processor board System flash (Read/Write)	
Rt99a# reload	//重新启动 IOS 的命令
Proceed with reload? [confirm]	//按【Enter】键便重新启动

⑤ IOS 受损后重启时的状况。

System Bootstrap, Version 12.1(3r)T2, RELEASE SOFTWARE (fc1)
Copyright (c) 2000 by cisco Systems, Inc.
Initializing memory for ECC
..
C2600 processor with 524288 Kbytes of main memory
Main memory is configured to 64 bit mode with ECC enabled
Readonly ROMMON initialized
Boot process failed...
The system is unable to boot automatically. The BOOT
environment variable needs to be set to a bootable image.
rommon 1 >

⑥ 恢复 IOS。

rommon 1 >　IP_ADDRESS=192.168.199.1
rommon 2 >　IP_SUBNET_MASK=255.255.255.0
rommon 3 >　DEFAULT_GATEWAY=192.168.199.1
rommon 4 >　TFTP_SERVER=192.168.199.254
rommon 5 >　TFTP_FILE=c2600-i-mz.122-28.bin
//必须先正确设置上面的 5 个变量，才能执行 tftpdnld 命令从 TFTP 服务器下载 IOS。
rommon 6 >　tftpdnld　　　　　　　//下载命令
IP_ADDRESS: 192.168.199.1
IP_SUBNET_MASK: 255.255.255.0
DEFAULT_GATEWAY: 192.168.199.1
TFTP_SERVER: 192.168.199.254
TFTP_FILE: c2600-i-mz.122-28.bin
Invoke this command for disaster recovery only.
WARNING: all existing data in all partitions on flash will be lost!
Do you wish to continue? y/n: [n]: y

```
.....[TIMED OUT]
TFTP: Operation terminated.
rommon 7 >
```

注意:在模拟器中恢复 IOS 不可能成功,但在真实的路由器上可以。

在模拟器中虽然不能用此方法恢复 IOS,但只要关闭模拟器,退出时单击【No】按钮,不保存所做修改,再重新打开模拟器就可恢复 IOS。前提是在做 IOS 恢复实验之前关闭模拟器,并选择保存所有修改。在真实的路由器上可用此方法恢复 IOS。

学习总结

本项目主要学习了路由器配置文件和 IOS 的备份与恢复。实现恢复的前提条件是事先已做好备份。备份文件一般都存放在 TFTP 服务器上。此外,配置文件也可复制、粘贴到文本文件中保存。

课后作业

完成上面的模拟实训,将实训过程的截图按顺序粘贴到一个 Word 文件里,并用适当的文字说明你对它的理解;总结本次实训所需要的主要命令及其作用,作为实训报告上交。

根据项目 01 中的路由器命名规则,务必把路由器命名为 RtIDa,其中 ID 为各自学号的最后两位数字。实训报告一律以 "ID 姓名项目号.doc" 命名,网络拓扑及其配置也以 "ID 姓名项目号.pkt" 为文件名保存并上交。例如,张三的 ID 为 03,他的文件名为 "03 张三 03.doc" 和 "03 张三 03.pkt"。

思考题

路由器的最主要功能是转发数据包,这需要查找路由表。管理员可以手工配置路由表项,这就是所谓的静态路由。那么静态路由怎么配置呢?

在下一个项目中我们将学习解决办法。

项目 04　串行线路与静态路由配置

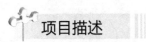

项目描述

若你是某单位的网络管理员，你需要构建连接两个园区的网络。每个园区网是一个以太网，两个园区网之间距离较远，由专线连接构成一个广域网。由于网络规模小、结构简单，故你决定采用静态路由。

网络拓扑

网络拓扑如图 4-1 所示。

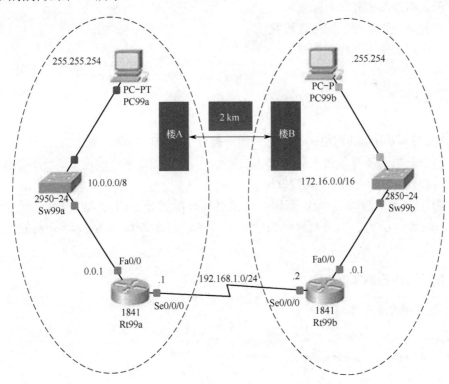

图 4-1　网络拓扑

学习目标

- 掌握串行线路和 HDLC 的配置方法；
- 掌握静态路由的配置方法；
- 掌握查看路由表的方法；
- 掌握验证配置是否正确的方法。

实训任务分解

① 配置路由器基本信息。
② 配置路由器串行接口。
③ 配置路由器以太网接口。
④ 配置 PC。
⑤ 测试网络连通性。
⑥ 查看路由表。
⑦ 配置静态路由并保存配置。
⑧ 查看路由表并测试网络连通性。
⑨ 改配默认路由并检查测试。

实训过程

构建如图 4-1 所示的网络：用串行专线连接两个远程以太网。请注意：

① 两台路由器都要增加一个 WIC-1T 模块，这样才能用串行线连接。当需要高速远距离传输数据时，可选用光纤及其接口模块。

② 为了表示子网的 IP 地址范围，路由器接口配置最小的 IP 地址，PC 配置最大的 IP 地址，或者，反之。如果交换机的接口足够多，则 Sw99a 和 Sw99b 分别能接很多台主机。

1. 配置路由器基本信息

（1）配置 Rt99a 基本信息

```
Router(config)# host Rt99a
Rt99a(config)# enable secret 99secret
Rt99a(config)# line con 0
Rt99a(config-line)# logg sync
Rt99a(config-line)# exec-timeout 0 0
```

Rt99a(config-line)# line vty 0 15
Rt99a(config-line)# password 99vty015
Rt99a(config-line)# login
Rt99a(config-line)# logg sync
Rt99a(config-line)# exec-timeout 0 0
Rt99a(config-line)# exit
Rt99a(config)# service password-encr
Rt99a(config)# no ip domain-lookup

(2) 配置 Rt99b 基本信息

Router(config)# host Rt99b
Rt99b(config)# enable secret 99secret
Rt99b(config)# line con 0
Rt99b(config-line)# logg sync
Rt99b(config-line)# exec-timeout 0 0
Rt99b(config-line)# line vty 0 15
Rt99b(config-line)# password 99vty015
Rt99b(config-line)# login
Rt99b(config-line)# logg sync
Rt99b(config-line)# exec-timeout 0 0
Rt99b(config-line)# exit
Rt99b(config)# service password-encr
Rt99b(config)# no ip domain-lookup

2．配置路由器串行接口

在配置串行接口时要注意，DCE 端必须配置时钟频率，而 DTE 端不用配置时钟频率。可通过查看串行控制器来确认是 DCE 端还是 DTE 端，命令用法如下：

Rt99a#　show　controller　se0/0/0

所有 Cisco 路由器串行接口的默认封装协议都是 HDLC，即以下命令已默认配置，故可省略：

Rt99a(config-if)# encapsulation　hdlc

(1) 查看 Rt99a 串行控制器

Rt99a# show controllers se0/0/0
Interface Serial0/1/0
Hardware is PowerQUICC MPC860
DCE V.35, clock rate 500000   //注意这行信息
idb at 0x81081AC4, driver data structure at 0x81084AC0
SCC Registers:
General [GSMR]=0x2:0x00000000, Protocol-specific [PSMR]=0x8
Events [SCCE]=0x0000, Mask [SCCM]=0x0000, Status [SCCS]=0x00
Transmit on Demand [TODR]=0x0, Data Sync [DSR]=0x7E7E
Interrupt Registers:

Config [CICR]=0x00367F80, Pending [CIPR]=0x0000C000
Mask   [CIMR]=0x00200000, In-srv   [CISR]=0x00000000
Command register [CR]=0x580
Port A [PADIR]=0x1030, [PAPAR]=0xFFFF
       [PAODR]=0x0010, [PADAT]=0xCBFF
Port B [PBDIR]=0x09C0F, [PBPAR]=0x0800E
       [PBODR]=0x00000, [PBDAT]=0x3FFFD
Port C [PCDIR]=0x00C, [PCPAR]=0x200
       [PCSO]=0xC20,   [PCDAT]=0xDF2, [PCINT]=0x00F
Receive Ring
       rmd(68012830): status 9000 length 60C address 3B6DAC4
       rmd(68012838): status B000 length 60C address 3B6D444
Transmit Ring
 --More--   //键入 q 退出

（2）查看 Rt99b 串行控制器

Rt99b# show controller se0/0/0
Interface Serial0/1/0
Hardware is PowerQUICC MPC860
DTE V.35 TX and RX clocks detected   //注意这行信息
idb at 0x81081AC4, driver data structure at 0x81084AC0
SCC Registers:
General [GSMR]=0x2:0x00000000, Protocol-specific [PSMR]=0x8
Events [SCCE]=0x0000, Mask [SCCM]=0x0000, Status [SCCS]=0x00
Transmit on Demand [TODR]=0x0, Data Sync [DSR]=0x7E7E
Interrupt Registers:
Config [CICR]=0x00367F80, Pending [CIPR]=0x0000C000
Mask   [CIMR]=0x00200000, In-srv   [CISR]=0x00000000
Command register [CR]=0x580
Port A [PADIR]=0x1030, [PAPAR]=0xFFFF
       [PAODR]=0x0010, [PADAT]=0xCBFF
Port B [PBDIR]=0x09C0F, [PBPAR]=0x0800E
       [PBODR]=0x00000, [PBDAT]=0x3FFFD
Port C [PCDIR]=0x00C, [PCPAR]=0x200
       [PCSO]=0xC20,   [PCDAT]=0xDF2, [PCINT]=0x00F
Receive Ring
       rmd(68012830): status 9000 length 60C address 3B6DAC4
       rmd(68012838): status B000 length 60C address 3B6D444
Transmit Ring
 --More--   //键入 q 退出

（3）配置 Rt99a 串行接口

时钟频率的可选值可用"clock rate ?"查询，实际应与线路的传输速率匹配。如 64 kbps 的专线选 64000。串行接口默认的封装协议为 HDLC。

Rt99a(config)# int se0/0/0	//指定串行接口
Rt99a(config-if)# desc WAN Link to Rt99b	//增加接口描述
Rt99a(config-if)# ip addr 192.168.1.1 255.255.255.0	//配置 IP 地址
Rt99a(config-if)# clock rate 500000	//在 DCE 端配置时钟频率
Rt99a(config-if)# no shut	

（4）配置 Rt99b 串行接口

注意：DTE 端不用配置时钟频率。串行接口默认的封装协议为 HDLC。

Rt99b(config)# int se0/0/0	//指定串行接口
Rt99b(config-if)# desc WAN Link to Rt99a	//增加接口描述
Rt99b(config-if)# ip addr 192.168.1.2 255.255.255.0	//配置 IP 地址
Rt99b(config-if)# no shut	

（5）测试 WAN 的连通性

Rt99a# ping 192.168.1.2	//检验配置是否正确
!!!!!	
Rt99b# ping 192.168.1.1	//检验配置是否正确
!!!!!	

3. 配置路由器以太网接口

（1）配置 Rt99a 网络接口

```
Rt99a(config)# int fa0/0
Rt99a(config-if)# desc LAN Link to Sw99a
Rt99a(config-if)# ip addr 10.0.0.1 255.0.0.0
Rt99a(config-if)# no shut
```

（2）配置 Rt99b 网络接口

```
Rt99b(config)# int fa0/0
Rt99b(config-if)# desc LAN Link to Sw99b
Rt99b(config-if)# ip addr 172.16.0.1 255.255.0.0
Rt99b(config-if)# no shut
Rt99b(config-if)# end
```

4. 配置 PC

在各台 PC 上配置 IP 地址、子网掩码和默认网关，然后单击【保存】按钮，保存已有配置。

① PC99a 的配置。
- IP 地址：10.255.255.254；
- 子网掩码：255.0.0.0；
- 默认网关：10.0.0.1。

② PC99b 的配置。
- IP 地址：172.16.255.254；
- 子网掩码：255.255.0.0；

- 默认网关：172.16.0.1。

5．测试网络连通性

（1）在 Rt99a 上测试网络连通性

```
Rt99a# ping 192.168.1.2                            //通
!!!!!
Rt99a# ping 10.255.255.254                         //通
!!!!!
Rt99a# ping 172.16.255.254                         //不通
.....
```

（2）在 Rt99b 上测试网络连通性

```
Rt99b# ping 192.168.1.1                            //通
!!!!!
Rt99b# ping 172.16.255.254                         //通
!!!!!
Rt99b# ping 10.255.255.254                         //不通
.....
```

为什么不通呢？

从以上测试结果看，两台路由器间是通的，Rt99a 到 PC99a 是通的，Rt99b 到 PC99b 也是通的，但是，从 Rt99a 到 PC99b 是不通的，从 Rt99b 到 PC99a 也是不通的。可以肯定 PC99a 与 PC99b 间也是不通的。

这是为什么呢？这是因为缺少路由配置。

PC99a 在 Rt99a 的直连网络上，而 PC99b 不在 Rt99a 的直连网络上，又没有到达 PC99b 所在网络的路由；PC99b 在 Rt99b 的直连网络上，而 PC99a 不在 Rt99b 的直连网络上，又没有到达 PC99a 所在网络的路由。

在配置静态路由前，先将网络拓扑和配置保存为文件名为 P04001.pkt 的文件备用，然后再将此文件另存为文件名为 P04002.pkt 的文件。

6．查看路由表

（1）查看 Rt99a 的路由信息

```
Rt99a# show ip route
Codes: C - connected, S - static, I - IGRP, R - RIP, M - mobile, B - BGP
       D - EIGRP, EX - EIGRP external, O - OSPF, IA - OSPF inter area
       N1 - OSPF NSSA external type 1, N2 - OSPF NSSA external type 2
       E1 - OSPF external type 1, E2 - OSPF external type 2, E - EGP
       i - IS-IS, L1 - IS-IS level-1, L2 - IS-IS level-2, ia - IS-IS inter area * - candidate default, U - per-user static route, o - ODR
       P - periodic downloaded static route
                                                    //以后一律省略此框内的信息
Gateway of last resort is not set
```

项目 04　串行线路与静态路由配置

```
C    10.0.0.0/8 is directly connected, FastEthernet0/0
C    192.168.1.0/24 is directly connected, Serial0/1/0
//以上 2 行显示只有直连路由，缺少到网络 172.16.0.0/16 的路由
```

（2）查看 Rt99b 的路由信息

```
Rt99b# show ip route
Gateway of last resort is not set
C    172.16.0.0/16 is directly connected, FastEthernet0/0
C    192.168.1.0/24 is directly connected, Serial0/1/0
//以上 2 行显示只有直连路由，缺少到网络 10.0.0.0/8 的路由
```

每次执行 show ip route 命令时都会先显示路由代码信息，由于该信息都是固定不变的，因此，为了节省篇幅，从此往后一律省略。

7. 配置静态路由并保存配置

（1）在 Rt99a 上配置静态路由并保存配置

```
Rt99a(config)# ip route 172.16.0.0 255.255.0.0 192.168.1.2    //配置静态路由
Rt99a(config)# end
Rt99a# write                                                   //保存配置
Building configuration...
[OK]                                                           //保存完毕
```

（2）在 Rt99b 上配置静态路由并保存配置

```
Rt99b(config)# ip route 10.0.0.0 255.0.0.0 192.168.1.1         //配置静态路由
Rt99b(config)# end
Rt99b# write                                                   //保存配置
Building configuration...
[OK]                                                           //保存完毕
```

8. 查看路由表并测试网络连通性

（1）再查看 Rt99a 路由信息

```
Rt99a# show ip route
Gateway of last resort is not set
C    10.0.0.0/8 is directly connected, FastEthernet0/0
S    172.16.0.0/16 [1/0] via 192.168.1.2
C    192.168.1.0/24 is directly connected, Serial0/1/0
//以上 3 行显示已增加了到网络 172.16.0.0/16 的静态路由
```

（2）再查看 Rt99b 路由信息

```
Rt99b#show ip route
Gateway of last resort is not set
S    10.0.0.0/8 [1/0] via 192.168.1.1
C    172.16.0.0/16 is directly connected, FastEthernet0/0
```

35

```
C    192.168.1.0/24 is directly connected, Serial0/1/0
```
//以上 3 行显示已增加了到网络 10.0.0.0/8 的静态路由

（3）再在 Rt99a 上测试网络连通性

```
Rt99a#ping 192.168.1.2                          //通
!!!!!
Rt99a#ping 10.255.255.254                       //通
!!!!!
Rt99a#ping 172.16.255.254                       //通
!!!!!
```

（4）再在 Rt99b 上测试网络连通性

```
Rt99b#ping 192.168.1.1                          //通
!!!!!
Rt99b#ping 172.16.255.254                       //通
!!!!!
Rt99b#ping 10.255.255.254                       //通
!!!!!
```

9．改配默认路由并检查测试

在配置默认路由前，先将拓扑网络和配置保存为文件名为 P04002.pkt 的文件备用，然后再另存为文件名为 P04003.pkt 的文件。

默认路由是指路由器在路由表中找不到到达目的网络的具体路由时最后会采用的路由。在所有的路由中，默认路由的优先级是最低的。

默认路由是静态路由的特例，通常在只有一个出口的末节网络（Stub Network）中使用，可以减少路由表项。在图 4-1 中，路由器 Rt99a 到达非直连网络的路径只有一条，那就是通过路由器 Rt99b，因此只要配置一条默认路由即可；同样，路由器 Rt99b 到达非直连网络的路径也只有一条，那就是通过路由器 Rt99a，因此也只要配置一条默认路由即可。

配置默认路由的命令如下：
```
ip route 0.0.0.0 0.0.0.0 { 网关地址 | 接口 }
```
要取消原来的配置，只需在原配置命令前加 no。例如，取消原来配置的静态路由：
```
Rt99b(config)# no ip route 10.0.0.0 255.0.0.0 192.168.1.1
```

（1）在 Rt99a 上改配默认路由

```
Rt99a(config)# no ip route 172.16.0.0 255.255.0.0 192.168.1.2    //取消原有配置
Rt99a(config)# ip route 0.0.0.0 0.0.0.0 192.168.1.2              //配置默认路由
Rt99a(config)# end
```

（2）在 Rt99b 上改配默认路由

```
Rt99b(config)# no ip route 10.0.0.0 255.0.0.0 192.168.1.1        //取消原有配置
Rt99b(config)# ip route 0.0.0.0 0.0.0.0 192.168.1.1              //配置默认路由
Rt99b(config)# end
```

项目 04　串行线路与静态路由配置

（3）查看 Rt99a 路由信息

```
Rt99a# show ip route
Gateway of last resort is 192.168.1.2 to network 0.0.0.0
C    10.0.0.0/8 is directly connected, FastEthernet0/0
C    192.168.1.0/24 is directly connected, Serial0/1/0
S*   0.0.0.0/0 [1/0] via 192.168.1.2                      //默认路由
```

注意默认路由与静态路由区别。

（4）查看 Rt99b 路由信息

```
Rt99b# show ip route
Gateway of last resort is 192.168.1.1 to network 0.0.0.0
C    172.16.0.0/16 is directly connected, FastEthernet0/0
C    192.168.1.0/24 is directly connected, Serial0/1/0
S*   0.0.0.0/0 [1/0] via 192.168.1.1                      //默认路由
```

注意默认路由与静态路由区别。

（5）在 Rt99a 上测试网络连通性

```
Rt99a# ping 192.168.1.2                                   //通
!!!!!
Rt99a# ping 10.255.255.254                                //通
.!!!!
Rt99a# ping 172.16.255.254                                //通
.!!!!
```

（6）在 Rt99b 测试网络连通性

```
Rt99b# ping 192.168.1.1                                   //通
!!!!!
Rt99b# ping 172.16.255.254                                //通
!!!!!
Rt99b# ping 10.255.255.254                                //通
!!!!!
```

（7）保存 Rt99a 的配置

```
Rt99a# copy run start
```

（8）保存 Rt99b 的配置

```
Rt99b# write
```

最后，在退出时，将网络拓扑和配置保存为文件名为 P04003.pkt 的文件。

学习总结

本项目的重点是串行线路和静态路由（默认路由）的配置。

串行线路一般用于连接远程网络。串行接口的 DCE 端必须配置时钟频率，而 DTE 端不用配置。查看串行控制器和配置时钟频率的命令如下：

Rt99a#　show　controller　se0/0/0
Rt99a(config-if)# clock　rate　64000

当 Cisco 路由器与其他厂商的设备串行连接时，一般还要配置 PPP：

Rt99a(config-if)# encapsulation　ppp

在本实训中，三个网络的 IP 地址和子网掩码分别为

- 10.0.0.0/255.0.0.0（简记为 10.0.0.0/8）；
- 172.16.0.0/255.255.0.0（简记为 172.16.0.0/16）；
- 192.168.1.0/255.255.255.0（简记为 192.168.1.0/24）。

将这三个网络连通的设备是路由器。路由器在转发数据包时需要正确的路由信息。管理员手工增加的路由称为静态路由，其配置命令如下：

Rt99a(config)# ip route　172.16.0.0　255.255.0.0　192.168.1.2
Rt99b(config)# ip route　10.0.0.0　255.0.0.0　192.168.1.1
Rt99b(config)# ip route　0.0.0.0　0.0.0.0　192.168.1.1　　　　　　//配置默认路由

网络号 0.0.0.0 配合子网掩码 0.0.0.0 表示任何网络。

课后作业

完成上面的模拟实训，将实训过程的截图按顺序粘贴到一个 Word 文件里并用适当的文字说明你对它的理解；总结本次实训所需要的主要命令及其作用，作为实训报告上交。

根据项目 01 中的路由器命名规则，务必把路由器命名为 RtIDa，其中 ID 为各自学号的最后两位数字。

实训报告一律以"ID 姓名项目号.doc"命名，网络拓扑及其配置也以"ID 姓名项目号.pkt"为文件名保存并上交。例如，张三的 ID 为 03，他的文件名为"03 张三 04.doc"和"03 张三 04.pkt"。

思考题

串行线路实际只需要 2 个 IP 地址，但在本实训中整个 C 类网络的其他 252 个 IP 地址却被浪费了。怎样有效利用 IP 地址，减少或避免 IP 地址的浪费呢？

在下一个项目中我们将学习解决办法。

项目 05　子网掩码与子网划分

项目描述

若你是某单位的网络管理员，为了有效利用 IP 地址，减少或避免 IP 地址浪费，你需要对网络进行合理规划，并用子网掩码来恰当地划分子网。例如，对于图 5-1 所示的网络，楼 A、B 和 C 中分别有 90、20 和 40 台主机需要连网，但如果只有一个 C 类 IP 网络地址 192.168.199.0 可用，则怎么办呢？

网络拓扑

网络拓扑如图 5-1 所示。

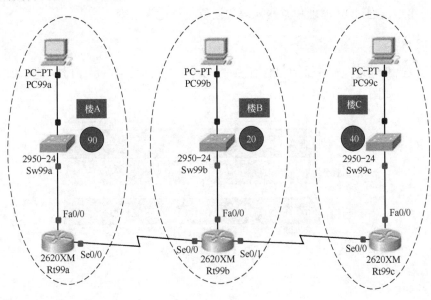

图 5-1　网络拓扑

学习目标

- 掌握子网掩码及其应用；

- 能计算子网数目和主机数目；
- 能确定子网地址、广播地址和主机地址；
- 能合理进行子网划分。

实训任务分解

① IP 地址与子网掩码。
② 子网划分。
③ 方案设计与选择。

实训过程

1．IP 地址与子网掩码

IP 地址和子网掩码的作用：
- IP 地址用于唯一地标识网络中的主机（节点）；
- 子网掩码用来识别某 IP 地址所属的网络；
- 路由器只根据网络地址来转发数据包；
- 网络地址通过 IP 地址和子网掩码的逐位相与运算获得。

（1）IP 地址

IPv4 中的 IP 地址是一个由 0 或 1 组成的 32 位二进制数字，用于唯一地标识网络中的主机（节点），如图 5-2 所示。

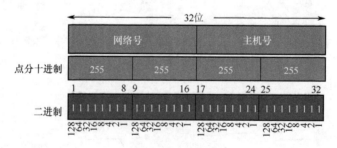

图 5-2　IP 地址

① 特殊 IP 地址。
- 127.0.0.1：本地回环（Loopback）测试地址；
- 255.255.255.255：全网（泛洪）广播地址；
- 0.0.0.0：代表任何网络；
- 主机号全为 1：代表该网络的所有主机，即该网络的广播地址，如 172.16.255.255。

② 私有 IP 地址。

项目 05 子网掩码与子网划分

- A 类：10.0.0.0～10.255.255.255；
- B 类：172.16.0.0～172.31.255.255；
- C 类：192.168.0.0～192.168.255.255。

利用私有地址可以节省 IP 地址，但私有地址不能在 Internet 上使用。

（2）子网掩码

子网掩码是一个 32 位的二进制码，用于识别 32 位的 IP 地址中的子网号和主机号。子网掩码总是与 IP 地址配对使用的。子网掩码左边连续为 1 的位对应于 IP 地址中的子网号，右边连续为 0 的位对应于 IP 地址中的主机号。通常所说的子网掩码的位数都是指左边连续为 1 的位数，而不是总的位数。本书在表述 IP 地址（如 172.16.0.0）和子网掩码（如 255.255.0.0）时，一般都简记为"IP 地址/子网掩码位数"（如 172.16.0.0/16）。

子网掩码各位的权值如图 5-3 所示，该图给出了一个八位组子网掩码的值。

```
权值：  128  64  32  16   8   4   2   1
         ↓   ↓   ↓   ↓   ↓   ↓   ↓   ↓
         1   0   0   0   0   0   0   0  =>  128
         1   1   0   0   0   0   0   0  =>  192
         1   1   1   0   0   0   0   0  =>  224
         1   1   1   1   0   0   0   0  =>  240
         1   1   1   1   1   0   0   0  =>  248
         1   1   1   1   1   1   0   0  =>  252
         1   1   1   1   1   1   1   0  =>  254
         1   1   1   1   1   1   1   1  =>  255
```

图 5-3 一个八位组子网掩码的值

网络掩码（Network Mask）即默认子网掩码（不分子网的掩码）。不需要进行子网划分的网络使用如图 5-4 所示的默认子网掩码或标准子网掩码。

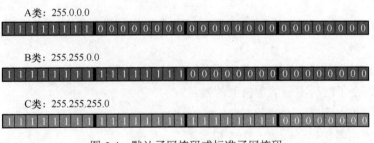

图 5-4 默认子网掩码或标准子网掩码

IP 地址中与子网掩码右边连续的 0 对应的位组成主机号。主机号二进制位全为 0 的 IP 地址是网络地址；主机号二进制位全为 1 的 IP 地址是广播地址。因此，如果主机号的位数为 N，则该子网可用的主机地址数为 2^N-2。例如，对于 26 位的子网掩码，主机号的位数为 6，则该子网可用的主机地址数为 $2^6-2=62$。由于主机号至少需要 2 位，因此，最长的子网掩码是 30 位（即 255.255.255.252）。

2．子网划分

子网划分的核心思想是"借用"主机号来"制造"新的"网络"。网络号的位数增加，则主机号的位数减少。为了节约有限的 IP 地址，所有类别的网络可以划分为更小的子网，这是通过子网掩码来实现的。

（1）子网划分要解答的五个问题

① 需要分割多少个子网？或选择多长的子网掩码？
② 最大的子网需要多少个主机地址？或每个子网要有多少个主机地址？
③ 有效的子网有哪些？
④ 每个子网的广播地址是什么？
⑤ 每个子网的有效主机地址有哪些？

注：把主机改称为节点更合适。

（2）子网划分及其计算方法

① 若子网掩码的长度减去标准掩码的长度为 s，则它将产生（2 的 s 次方）个子网，子网数目=2^s。
② 若 32 减去子网掩码的长度为 t，则每个子网最多有（2 的 t 次方−2）个主机地址，主机数目=2^t-2。
③ 块大小=256−十进制的子网掩码，有效子网号=0+块大小的倍数（从 0 到十进制的子网掩码）。
④ 每个子网的广播地址=下个子网号−1。
⑤ 除去子网内全为 0 和全为 1 的地址剩下的就是子网内的有效主机地址。

（3）创建子网的步骤

① 确定所需要的子网号的数目：
- 每个子网必须有一个子网号；
- 每个广域网连接必须有一个子网号。

② 确定每个子网所需要的主机号的数目：
- 每个主机必须有一个主机号；
- 每个路由器接口必须有一个主机号。

③ 根据以上需求，确定：
- 一个用于整个网络的子网掩码；
- 每个物理网段一个唯一的子网号；
- 每个子网的主机号范围。

（4）C 类网络子网划分举例

[例 1]　若 C 类网络地址为 192.168.199.0，子网掩码为 26 位，即 255.255.255.192 (/26)，则：

① 该网络可以划分多少个子网？
② 每个子网的主机数目是多少？
③ 每个子网的 IP 地址是什么？
④ 每个子网的广播地址是什么？
⑤ 每个子网的有效主机地址范围是什么？

解答：

① $s=26-24=2$，子网数$=2^2=4$。

② $t=32-26=6$，主机数$=2^6-2=62$。

③ 块大小$=256-192=64$，所以四个子网的 IP 地址分别是：

- 192.168.199.0；
- 192.168.199.64；
- 192.168.199.128；
- 192.168.199.192。

④ 广播地址=下个子网地址-1，分别是：

- 192.168.199.63；
- 192.168.199.127；
- 192.168.199.191；
- 192.168.199.255。

⑤ 四个子网的有效主机地址范围分别是：

- 192.168.199.1～192.168.199.62；
- 192.168.199.65～192.168.199.126；
- 192.168.199.129～192.168.199.190；
- 192.168.199.193～192.168.199.254。

[例 2] 若某节点的 IP 地址为 192.168.199.35，子网掩码为 255.255.255.224，则该节点的子网地址和广播地址是什么？

解答：

① 计算块大小：256-224=32。

② 求子网号：0，32，64，…

35 介于 32 与 64 之间，所以该节点的子网地址是 192.168.199.32，广播地址是 192.168.199.63。

[例 3] 若某节点的 IP 地址为 192.168.199.18，子网掩码为 255.255.255.252，则该节点的子网地址和广播地址是什么？

解答：

① 计算块大小：256-252=4。

② 求子网号：0，4，8，12，16，20，…

18 介于 16 与 20 之间，所以该节点的子网地址是 192.168.10.16，广播地址是 192.168.10.19。

30位的子网掩码特别适合用于点对点的连接，因为它恰好有2个可用的主机地址。

（5）B类网络子网划分举例

[例1]　若B类网络地址为172.16.0.0，子网掩码为19位，即255.255.224.0(/19)，则：
① 该网络可以划分多少个子网？
② 每个子网的主机数目是多少？
③ 每个子网的IP地址是什么？
④ 每个子网的广播地址是什么？
⑤ 每个子网的有效主机地址范围是什么？

解答：

① s=19−16=3，子网数=2^3=8。

② t=32−19=13，主机数=2^{13}−2=8190。

③ 块大小=256−224=32，所以八个子网的IP地址分别是：
- 172.16.0.0；
- 172.16.32.0；
- 172.16.64.0；
- 172.16.96.0；
- 172.16.128.0；
- 172.16.160.0；
- 172.16.192.0；
- 172.16.224.0。

④ 广播地址=下个子网地址−1，所以八个子网的广播地址分别是：
- 172.16.31.255；
- 172.16.63.255；
- 172.16.95.255；
- 172.16.127.255；
- 172.16.159.255；
- 172.16.191.255；
- 172.16.223.255；
- 172.16.255.255。

⑤ 八个子网的有效主机地址范围分别是：
- 172.16.0.1～172.16.31.254；
- 172.16.32.1～172.16.63.254；
- 172.16.64.1～172.16.95.254；
- 172.16.96.1～172.16.127.254；
- 172.16.128.1～172.16.159.254；

- 172.16.160.1～172.16.191.254；
- 172.16.192.1～172.16.223.254；
- 172.16.224.1～172.16.255.254。

[例 2] 若某节点的 IP 地址和子网掩码为 172.16.199.255/20，则该节点的子网地址和广播地址是什么？该子网的主机号范围是什么？

解答：

① /20 表示 255.255.240.0，块大小为 256-240=16。

② 子网号：0，16，32，48，64，80，96，128，160，192，224，…

199 介于 192 与 224 之间，所以该节点的子网地址是 172.16.192.0，广播地址是 172.16.223.255。

③ 该子网的主机号范围是：192.1～112.254。

[例 3] 若某节点的 IP 地址为 172.16.199.13，子网掩码为 255.255.255.252，则该节点的子网地址和广播地址是什么？

解答：

① 计算块大小：256-252=4。

② 求子网号：0，4，8，12，16，20，…

13 介于 12 与 16 之间，所以该节点的子网地址是 172.16.199.12，广播地址是 172.16.199.15。

（6）A 类网络子网划分举例

[例 1] 若某节点的 IP 地址和子网掩码为 10.199.66.33/20，则该节点的子网地址和广播地址是什么？该子网的有效主机地址的范围是什么？

解答：

① /20 表示 255.255.240.0，块大小为 256-240=16。

② 子网号：0，16，32，48，64，80，…

66 介于 64 与 80 之间，所以该节点的子网地址是 10.199.64.0，广播地址是 10.199.79.255。

③ 该子网的有效主机地址的范围是：10.199.64.1～10.199.79.254。

[例 2] 若某节点的 IP 地址和子网掩码为 10.199.66.25/30，则该节点的子网和广播地址是什么？该子网的有效主机地址是什么？

解答：

① /30 表示 255.255.255.252，256-252=4。

② 子网号：0，4，8，12，16，20，24，28，32。

25 介于 24 与 28 之间，所以该节点的子网地址是 10.199.66.24，广播地址是 10.199.66.27。

该子网的有效主机地址是：10.199.66.25 和 10.199.66.26。

3．方案设计与选择

单一子网掩码还不能解决开头提出的问题，还必须采用变长子网掩码，将子网进一步划分为更小的子网。

（1）变长子网掩码（VLSM）

变长子网掩码左边连续为 1 的位数可以是任意多位。因此，它突破了有类网络的界限，产生了无类网络或超网的概念。

（2）方案设计

采用变长子网掩码，可将子网进一步划分为更小的子网。对于开头提出的问题，可以如下划分子网：

- 对于有 90 台主机的子网，块大小为 128，子网掩码为 255.255.255.128 (/25)。它将一个 C 类网络的地址空间一分为二，然后，对另一半再细分。
- 对于有 40 台主机的子网，块大小为 64，子网掩码为 255.255.255.192 (/26)。它将一个有 128 个 IP 地址的块一分为二，然后，对另一半可再细分。
- 对于有 20 台主机的子网，块大小为 32，子网掩码为 255.255.255.224 (/27)。它将一个有 64 个 IP 地址的块一分为二，然后，对另一半可再细分。
- 对于只有 2 台主机的子网，块大小为 4，子网掩码为 255.255.255.252 (/30)。它将一个有 32 个 IP 地址的块一分为八。

这样划分子网后，可以选用的方案就有多种，其中的两种分配方案如图 5-5 所示。

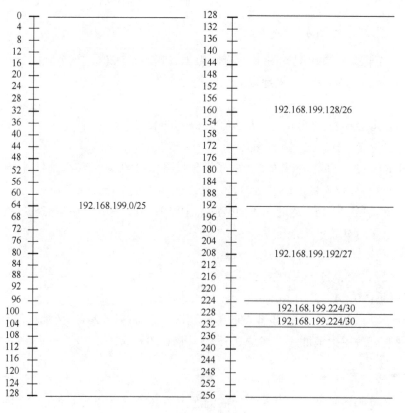

图 5-5 分配方案

项目 05　子网掩码与子网划分

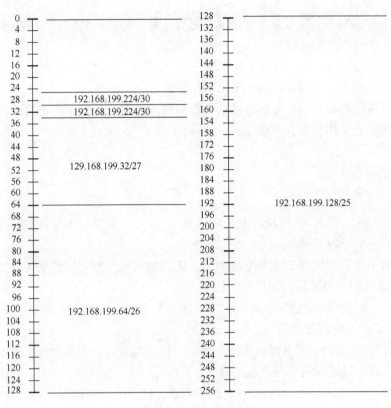

图 5-5　分配方案（续）

（3）方案选择

选用前一种分配方案，网络拓扑如图 5-6 所示，该图给出了具体的子网划分和 IP 地址分配情况，圆圈中的数字是各以太网的最大主机数，这显然符合需求，而且还有一定的余量。

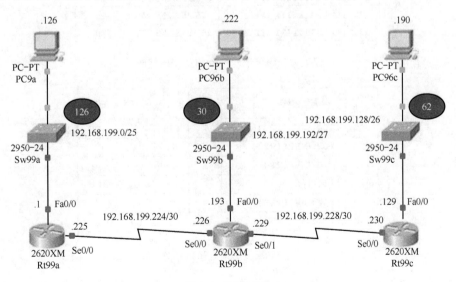

图 5-6　网络拓扑

47

本例采用 VLSM 后，不仅满足了需求，而且还有一定的扩展空间：192.168.199.232 ～ 192.168.199.255 的地址尚未分配。

注意：为路由器 Rt99a 和 Rt99c 各增配一块 WIC-1T 接口卡，为路由器 Rt99b 增配一块 WIC-2T 接口卡后，方可按图 5-6 连接串行线路。

请选择与图 5-6 中一致的路由器和交换机型号，完成如图 5-6 所示的连接后，将网络拓扑和配置保存为文件名为 P05001.pkt 的文件。

学习总结

为了高效利用有限的 IP 地址、减少或避免 IP 地址的浪费，所有类别的网络通过子网掩码可以划分为更小的子网。

子网掩码用来识别某 IP 地址所属的网络。路由器只根据网络地址来转发数据包。网络地址通过 IP 地址和子网掩码的逐位相与运算获得。

应用变长子网掩码（VLSM）可以更高效地利用有限的 IP 地址，可以打破传统的以类（Class）为标准的地址划分方法。

可用于 C 类网络的子网掩码如图 5-7 所示，C 类地址子网划分如图 5-8 所示，图 5-7 和图 5-8 是对 C 类地址子网划分情况的总结。

可用于C类地址的子网掩码		
二进制	点分十进制	CIDR记法
11111111 11111111 10000000	255.255.255.128	/25
11111111 11111111 11000000	255.255.255.192	/26
11111111 11111111 11100000	255.255.255.224	/27
11111111 11111111 11110000	255.255.255.240	/28
11111111 11111111 11111000	255.255.255.248	/29
11111111 11111111 11111100	255.255.255.252	/30

图 5-7　可用于 C 类网络的子网掩码

CIDR记法	子网号掩码	子网位数	主机位数	块大小	子网数	主机数
/25	10000000（128）	1	7	128	2	126
/26	11000000（192）	2	6	64	4	62
/27	11100000（224）	3	5	32	8	30
/28	11110000（240）	4	4	16	16	14
/29	11111000（248）	5	3	8	32	6
/30	11111100（252）	6	2	4	64	2

图 5-8　C 类地址子网划分

可用于 B 类网络的子网掩码如图 5-9 所示，该图是对 B 类地址子网划分情况的总结。

项目 05　子网掩码与子网划分

可用于 B 类地址的子网掩码	
255.255.128.0　（/17）	255.255.255.0　　（/24）
255.255.192.0　（/18）	255.255.255.128（/25）
255.255.224.0　（/19）	255.255.255.192（/26）
255.255.240.0　（/20）	255.255.255.224（/27）
255.255.248.0　（/21）	255.255.255.240（/28）
255.255.252.0　（/22）	255.255.255.248（/29）
255.255.254.0　（/23）	255.255.255.252（/30）

图 5-9　可用于 B 类网络的子网掩码

可用于 A 类网络的子网掩码如图 5-10 所示，该图是对 A 类地址子网划分情况的总结。

可用于 A 类地址的子网掩码	
255.128.0.0　　（/9）	255.255.240.0　　（/20）
255.192.0.0　　（/10）	255.255.248.0　　（/21）
255.224.0.0　　（/11）	255.255.252.0　　（/22）
255.240.0.0　　（/12）	255.255.254.0　　（/23）
255.248.0.0　　（/13）	255.255.255.0　　（/24）
255.252.0.0　　（/14）	255.255.255.128（/25）
255.254.0.0　　（/15）	255.255.255.192（/26）
255.255.0.0　　（/16）	255.255.255.224（/27）
255.255.128.0　（/17）	255.255.255.240（/28）
255.255.192.0　（/18）	255.255.255.248（/29）
255.255.224.0　（/19）	255.255.255.252（/30）

图 5-10　可用于 A 类网络的子网掩码

所有可用的子网掩码如图 5-11 所示。

子网掩码	CIDR 记法	子网掩码	CIDR 记法
255.0.0.0	/8	255.255.240.0	/20
255.128.0.0	/9	255.255.248.0	/21
255.192.0.0	/10	255.255.252.0	/22
255.224.0.0	/11	255.255.254.0	/23
255.240.0.0	/12	255.255.255.0	/24
255.248.0.0	/13	255.255.255.128	/25
255.252.0.0	/14	255.255.255.192	/26
255.254.0.0	/15	255.255.255.224	/27
255.255.0.0	/16	255.255.255.240	/28
255.255.128.0	/17	255.255.255.248	/29
255.255.192.0	/18	255.255.255.252	/30
255.255.224.0	/19		

图 5-11　所有可用的子网掩码

注意：变长子网掩码（VLSM）不受上面的限制。

课后作业

完成图 5-6 所示网络的连接，以"ID 姓名项目号.pkt"为文件名保存，为下一个项目做准备。例如，张三的 ID 为 03，他的文件名为"03 张三 05.pkt"。

思考题

在图 5-6 中的每台路由器上怎么配置静态路由？如何减少路由表项？

若想为图 5-6 中的串行线路各增加一条备份路径，则怎么配置静态路由呢？

在下一个项目中我们将学习有关的知识和技术。

项目 06 静态路由综合配置

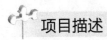 项目描述

你是某单位的网络管理员，因网络结构简单决定采用静态路由或默认路由；此外，为了提高两个网络之间通信的可靠性，你决定增加备份链路，并配置浮动静态路由或浮动默认路由，怎么做？

网络拓扑

网络拓扑如图 6-1 所示。

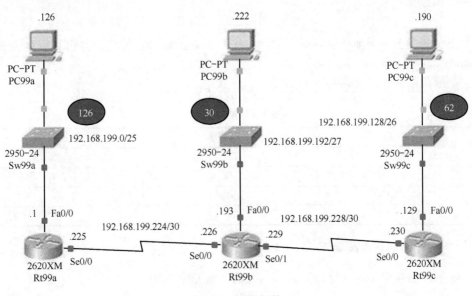

图 6-1 网络拓扑

学习目标

- 理解和掌握子网掩码及其应用；
- 重点掌握串行线路的配置方法；

51

- 重点掌握静态路由和默认路由的配置方法；
- 掌握浮动静态路由的配置方法；
- 掌握验证配置是否正确的方法。

实训任务分解

① 完成路由器基本配置。
② 完成路由器接口配置。
③ 完成 PC IP 地址配置。
④ 测试与检验网络。
⑤ 配置与检验静态路由。
⑥ 改配默认路由并检验。
⑦ 增加和配置备份链路。
⑧ 配置与检验浮动静态路由。
⑨ 模拟主链路中断与恢复。

实训过程

本次实训在上一个项目的基础上继续。先打开 P05001.pkt 文件，并将其另存为文件名为 P06001.pkt 的文件。

实训网络拓扑和配置数据分别如图 6-1 和表 6-1 所示。

表 6-1 配置数据

子网掩码	CIDR 记法	主 机 数
255.255.255.128	/25	126
255.255.255.192	/26	62
255.255.255.224	/27	30
255.255.255.252	/30	2

1. 完成路由器基本配置

在以下的配置中，每台路由器都不再开启远程登录功能。

（1）完成 Rt99a 的基本配置

```
Router(config)# host Rt99a
Rt99a(config)# no ip domain-lookup
Rt99a(config)# enable secret 99secret
Rt99a(config)# service password-encr
Rt99a(config)# line con 0
Rt99a(config-line)# password 99con0
```

```
Rt99a(config-line)# login
Rt99a(config-line)# logg sync
Rt99a(config-line)# exec-timeout 0 0
Rt99a(config-line)# end
```

（2）完成 Rt99b 的基本配置

```
Router(config)# host Rt99b
Rt99b(config)# no ip domain-lookup
Rt99b(config)# enable secret 99secret
Rt99b(config)# service password-encr
Rt99b(config)# line con 0
Rt99b(config-line)# password 99con0
Rt99b(config-line)# login
Rt99b(config-line)# logg sync
Rt99b(config-line)# exec-timeout 0 0
Rt99b(config-line)# end
```

（3）完成 Rt99c 的基本配置

```
Router(config)# host Rt99c
Rt99c(config)# no ip domain-lookup
Rt99c(config)# enable secret 99secret
Rt99c(config)# service password-encr
Rt99c(config)# line con 0
Rt99c(config-line)# password 99con0
Rt99c(config-line)# login
Rt99c(config-line)# logg sync
Rt99c(config-line)# exec-timeout 0 0
Rt99c(config-line)# end
```

（4）保存每台路由器的配置

```
Rt99a# write
Rt99b# write
Rt99c# write
```

完成以上配置后，先将相关配置保存到 P06001.pkt 文件中备用，再将其另存为文件名为 P06002.pkt 的文件，然后继续进行实训工作。

2．完成路由器接口配置

为以太网接口配置 IP 地址和子网掩码，为串行接口配置 IP 地址和子网掩码，在 DCE 端配置时钟频率。

（1）配置 Rt99a 的接口

```
Rt99a(config)# int fa0/0                    //指定配置的接口为 Fa0/0
Rt99a(config-if)# desc LAN link to Sw99a in building A
```

```
Rt99a(config-if)# ip addr 192.168.199.1 255.255.255.128
Rt99a(config-if)# no shut
Rt99a(config-if)# int se0/0                              //指定配置的接口为 Se0/0
Rt99a(config-if)# desc WAN link to Rt99b in building B
Rt99a(config-if)# ip addr 192.168.199.225 255.255.255.252
Rt99a(config-if)# clock rate 2000000                     //在 DCE 端配置时钟频率
Rt99a(config-if)# no shut
```

（2）配置 Rt99c 的接口

```
Rt99c(config)# int fa0/0                                 //指定配置的接口为 Fa0/0
Rt99c(config-if)# desc LAN link to Sw99c in building C
Rt99c(config-if)# ip addr 192.168.199.129 255.255.255.192
Rt99c(config-if)# no shut
Rt99c(config-if)# int se0/0                              //指定配置的接口为 Se0/0
Rt99c(config-if)# desc WAN link to Sw99b in building B
Rt99c(config-if)# ip addr 192.168.199.230 255.255.255.252
Rt99c(config-if)# clock rate 1000000                     //在 DCE 端配时钟频率
Rt99c(config-if)# no shut
```

（3）配置 Rt99b 的接口

```
Rt99b(config)# int fa0/0                                 //指定配置的接口为 Fa0/0
Rt99b(config-if)# desc LAN link to Sw99b in building B
Rt99b(config-if)# ip addr 192.168.199.193 255.255.255.224
Rt99b(config-if)# no shut

Rt99b(config-if)# int se0/0                              //指定配置的接口为 Se0/0
Rt99b(config-if)# desc WAN link to Rt99a in building A
Rt99b(config-if)# ip addr 192.168.199.226 255.255.255.252
Rt99b(config-if)# no shut

Rt99b(config-if)# int se0/1                              //指定配置的接口为 Se0/1
Rt99b(config-if)# desc WAN link to Rt99c in building C
Rt99b(config-if)# ip addr 192.168.199.229 255.255.255.252
Rt99b(config-if)# no shut
```

（4）保存每台路由器的配置

```
Rt99a# write
Rt99b# write
Rt99c# write
```

3．完成 PC IP 地址配置

在各台 PC 上按如下数据配置 IP 地址、子网掩码和默认网关。

① PC99a 的配置。
- IP 地址：192.168.199.126；

- 子网掩码：255.255.255.128；
- 默认网关：192.168.199.1。

② PC99b 的配置。
- IP 地址：192.168.199.222；
- 子网掩码：255.255.255.224；
- 默认网关：192.168.199.193。

③ PC99c 的配置。
- IP 地址：192.168.199.190；
- 子网掩码：255.255.255.192；
- 默认网关：192.168.199.129。

4．测试与检验网络

测试各 LAN 和 WAN 的连通性，以检验以上的配置是否正确。

（1）在 Rt99a 上测试网络连通性

Rt99a# ping 192.168.199.126	//测试到 PC99a 的连通性
.!!!!	
Rt99a# ping 192.168.199.226	//测试到 Rt99b 的连通性
!!!!!	

（2）在 Rt99c 上测试网络连通性

Rt99c# ping 192.168.199.190	//测试到 PC99c 的连通性
.!!!!	
Rt99c# ping 192.168.199.229	//测试到 Rt99b 的连通性
!!!!!	

（3）在 Rt99b 上测试网络连通性

Rt99b# ping 192.168.199.222	//测试到 PC99b 的连通性
!!!!!	
Rt99b# ping 192.168.199.225	//测试到 Rt99a 的连通性
!!!!!	
Rt99b# ping 192.168.199.230	//测试到 Rt99c 的连通性
!!!!!	

（4）在 PC99a 上测试网络连通性

C:\> ping 192.168.199.222	//测试到 PC99b 的连通性
……	
C:\> ping 192.168.199.190	//测试到 PC99c 的连通性
……	

（5）在 PC99b 上测试网络连通性

| C:\> ping 192.168.199.126 | //测试到 PC99a 的连通性 |
| …… | |

```
C:\> ping 192.168.199.190          //测试到 PC99c 的连通性
……
```

（6）在 PC99c 上测试网络连通性

```
C:\> ping 192.168.199.126          //测试到 PC99a 的连通性
……
C:\> ping 192.168.199.222          //测试到 PC99b 的连通性
……
```

5. 配置与检验静态路由

验证前面的配置都正确以后，单击【保存】按钮将相关存配置保存到 P06002.pkt 文件中备用，然后再把它另存为文件名为 P06003.pkt 的文件。接着在每台路由器上检查和配置静态路由。以下（4）、（5）、（6）是重点。

（1）查看 Rt99a 的路由表

```
Rt99a# show ip route
Gateway of last resort is not set
     192.168.199.0/24 is variably subnetted, 2 subnets, 2 masks
C       192.168.199.0/25 is directly connected, FastEthernet0/0
C       192.168.199.224/30 is directly connected, Serial0/0
//只有 2 条直连路由，但共有 5 个子网
```

（2）查看 Rt99b 的路由表

```
Rt99b# show ip route
Gateway of last resort is not set
     192.168.199.0/24 is variably subnetted, 3 subnets, 2 masks
C       192.168.199.192/27 is directly connected, FastEthernet0/0
C       192.168.199.224/30 is directly connected, Serial0/0
C       192.168.199.228/30 is directly connected, Serial0/1
//只有 3 条直连路由，但共有 5 个子网
```

（3）查看 Rt99c 的路由表

```
Rt99c# show ip route
Gateway of last resort is not set
     192.168.199.0/24 is variably subnetted, 2 subnets, 2 masks
C       192.168.199.128/26 is directly connected, FastEthernet0/0
C       192.168.199.228/30 is directly connected, Serial0/0
//只有 2 条直连路由，但共有 5 个子网
```

（4）在 Rt99a 上配置静态路由

```
Rt99a(config)# ip route 192.168.199.192 255.255.255.224 192.168.199.226
Rt99a(config)# ip route 192.168.199.228 255.255.255.252 192.168.199.226
Rt99a(config)# ip route 192.168.199.128 255.255.255.192 192.168.199.226
```

注意：3 条路由下一跳的地址都是 192.168.199.226。

（5）在 Rt99b 上配置静态路由

Rt99b(config)# ip route 192.168.199.0 255.255.255.128 192.168.199.225
Rt99b(config)# ip route 192.168.199.128 255.255.255.192 192.168.199.230

注意：静态路由的管理距离默认为 1。

（6）在 Rt99c 上配置静态路由

Rt99c(config)# ip route 192.168.199.192 255.255.255.224 192.168.199.229
Rt99c(config)# ip route 192.168.199.224 255.255.255.252 192.168.199.229
Rt99c(config)# ip route 192.168.199.0 255.255.255.128 192.168.199.229

注意：3 条路由的下一跳地址都是 192.168.199.229。

（7）查看 Rt99a 的路由表

```
Rt99a# show ip route
Gateway of last resort is not set
     192.168.199.0/24 is variably subnetted, 5 subnets, 4 masks
C       192.168.199.0/25 is directly connected, FastEthernet0/0
S       192.168.199.128/26 [1/0] via 192.168.199.226
S       192.168.199.192/27 [1/0] via 192.168.199.226
C       192.168.199.224/30 is directly connected, Serial0/0
S       192.168.199.228/30 [1/0] via 192.168.199.226
//有 2 条直连路由和 3 条静态路由（共有 5 个子网）
```

（8）查看 Rt99b 的路由表

```
Rt99b# show ip route
Gateway of last resort is not set
     192.168.199.0/24 is variably subnetted, 5 subnets, 4 masks
S       192.168.199.0/25 [1/0] via 192.168.199.225
S       192.168.199.128/26 [1/0] via 192.168.199.230
C       192.168.199.192/27 is directly connected, FastEthernet0/0
C       192.168.199.224/30 is directly connected, Serial0/0
C       192.168.199.228/30 is directly connected, Serial0/1
//有 3 条直连路由和 2 条静态路由（共有 5 个子网）
```

（9）查看 Rt99c 的路由表

```
Rt99c# show ip route
Gateway of last resort is not set
     192.168.199.0/24 is variably subnetted, 5 subnets, 4 masks
S       192.168.199.0/25 [1/0] via 192.168.199.229
C       192.168.199.128/26 is directly connected, FastEthernet0/0
S       192.168.199.192/27 [1/0] via 192.168.199.229
S       192.168.199.224/30 [1/0] via 192.168.199.229
C       192.168.199.228/30 is directly connected, Serial0/0
//有 2 条直连路由和 3 条静态路由（共有 5 个子网）
```

（10）在 PC99a 上测试网络连通性

C:\> ping 192.168.199.222   //测试到 PC99b 的连通性

```
!!!!!
C:\> ping 192.168.199.190          //测试到 PC99c 的连通性
!!!!!
```

（11）在 PC99b 上测试网络连通性

```
C:\> ping 192.168.199.126          //测试到 PC99a 的连通性
!!!!!
C:\> ping 192.168.199.190          //测试到 PC99c 的连通性
!!!!!
```

（12）在 PC99c 上测试网络连通性

```
C:\> ping 192.168.199.126          //测试到 PC99a 的连通性
!!!!!
C:\> ping 192.168.199.222          //测试到 PC99b 的连通性
!!!!!
```

在各台 PC 上重新测试的结果验证了上面的配置是正确的，网络全通了。

（13）保存每台路由器的配置

```
Rt99a# write
Rt99b# write
Rt99c# write
```

6. 改配默认路由并检验

单击【保存】按钮将相关配置保存到 P06003.pkt 文件中备用，再将其另存为文件名为 P06004.pkt 的文件，然后继续进行实训工作。

Rt99a 和 Rt99c 都把目标网络不是其直连网络的数据包转发给 Rt99b，因此，可将具体的静态路由改为默认路由以减少路由表项。

（1）在 Rt99a 上改配默认路由

```
Rt99a(config)# no ip route 192.168.199.192 255.255.255.224 192.168.199.226
Rt99a(config)# no ip route 192.168.199.228 255.255.255.252 192.168.199.226
Rt99a(config)# no ip route 192.168.199.128 255.255.255.192 192.168.199.226
//以上 3 行命令取消了原先配置的 3 条静态路由
Rt99a(config)# ip route 0.0.0.0 0.0.0.0 192.168.199.226         //配置默认路由
Rt99a(config)# end
```

（2）在 Rt99c 上改配默认路由

```
Rt99c(config)# no ip route 192.168.199.192 255.255.255.224 192.168.199.229
Rt99c(config)# no ip route 192.168.199.224 255.255.255.252 192.168.199.229
Rt99c(config)# no ip route 192.168.199.0 255.255.255.128 192.168.199.229
//以上 3 行命令取消了原先配置的 3 条静态路由
Rt99c(config)# ip route 0.0.0.0 0.0.0.0 192.168.199.229         //配置默认路由
Rt99c(config)# end
```

（3）查看 Rt99a 的路由表

```
Rt99a# show ip route
```

```
Gateway of last resort is 192.168.199.226 to network 0.0.0.0        //注意这一行的变动
       192.168.199.0/24 is variably subnetted, 2 subnets, 2 masks
C      192.168.199.0/25 is directly connected, FastEthernet0/0
C      192.168.199.224/30 is directly connected, Serial0/0/0
S*     0.0.0.0/0 [1/0] via 192.168.199.226
//以上 3 行显示有 2 条直连路由和 1 条默认路由（共有 5 个子网）
```

（4）查看 Rt99c 的路由表

```
Rt99c# show ip route
Gateway of last resort is 192.168.199.229 to network 0.0.0.0        //注意这一行的变动
       192.168.199.0/24 is variably subnetted, 2 subnets, 2 masks
C      192.168.199.128/26 is directly connected, FastEthernet0/0
C      192.168.199.228/30 is directly connected, Serial0/0/0
S*     0.0.0.0/0 [1/0] via 192.168.199.229
//有 2 条直连路由和 1 条默认路由（共有 5 个子网）
```

在图 6-1 中，路由器 Rt99a 到达非直连网络（路由器 Rt99c）的路径只有一条，那就是通过路由器 Rt99b，因此只要配置一条默认路由即可。

同样，路由器 Rt99c 到达非直连网络（路由器 Rt99a）的路径也只有一条，那就是通过路由器 Rt99b，因此也只要配置一条默认路由即可。

通过在 Rt99a 和 Rt99c 上配置默认路由，使它们的路由表项各自从 5 项减到了 3 项。如果网络的规模增大，则路由表项减少的幅度将更大。

请注意，在完成配置前后，如图 6-2 所示的网络拓扑在模拟器中有何不同？

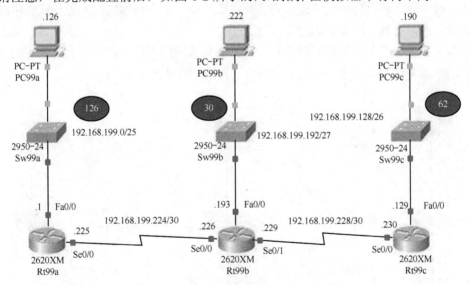

图 6-2 网络拓扑

（5）在 PC99a 上测试网络连通性

```
C:\> ping 192.168.199.222
!!!!!
C:\> ping 192.168.199.190
!!!!!
```

（6）在 PC99b 上测试网络连通性

C:\> ping 192.168.199.126
!!!!!
C:\> ping 192.168.199.190
!!!!!

（7）在 PC99c 上测试网络连通性

C:\> ping 192.168.199.126
!!!!!
C:\> ping 192.168.199.222
!!!!!

（5）～（7）的测试结果表明，整个网络已经全部连通了。

（8）保存每台路由器的配置

Rt99a# write
Rt99b# write
Rt99c# write

单击【保存】按钮将相关配置保存到 P06004.pkt 文件中备用，再将其另存为文件名为 P06005.pkt 的文件，然后继续进行实训工作。

7．增加和配置备份链路

两个网络之间若只有一条链路，则当其出现故障时两个网络之间的通信就会中断。为了增加可靠性，可以增加备份链路。

在图 6-2 中，任何一条串行线出现故障都将造成网络不能全部连通。若想提高网络的可靠性，可为串行线路各增加一条备份链路，网络拓扑如图 6-3 所示。

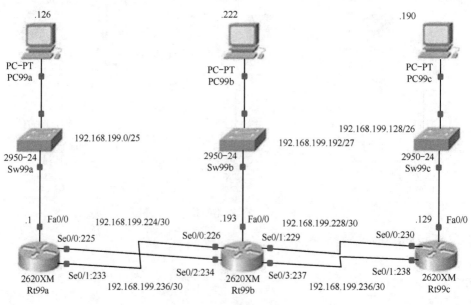

图 6-3　网络拓扑

项目 06　静态路由综合配置

在路由器 Rt99a 和 Rt99c 上各增配一块 WIC-1T 接口卡，在路由器 Rt99b 上增配一块 WIC-2T 接口卡后，方可按图 6-3 所示网络拓扑连接备份链路。

（1）在 Rt99a 上增配一块 WIC-1T 接口卡（如图 6-4 所示）

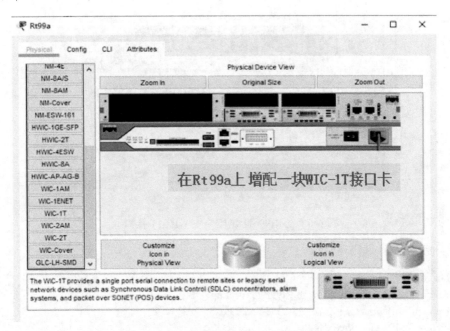

图 6-4　在 Rt99a 上增配一块 WIC-1T 接口卡

（2）在 Rt99b 上增配一块 WIC-2T 接口卡（如图 6-5 所示）

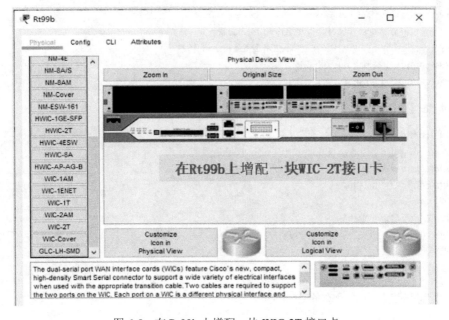

图 6-5　在 Rt99b 上增配一块 WIC-2T 接口卡

（3）在 Rt99c 上增配一块 WIC-1T 接口卡（如图 6-6 所示）

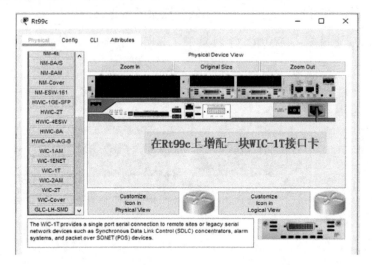

图 6-6　在 Rt99c 上增配一块 WIC-1T 接口卡

（4）配置 Rt99a 增加的接口

Rt99a(config)# int se0/1
Rt99a(config-if)# desc WAN link to Rt99b for backup
Rt99a(config-if)# ip addr 192.168.199.233 255.255.255.252
Rt99a(config-if)# clock rate 64000
Rt99a(config-if)# no shut

（5）配置 Rt99c 增加的接口

Rt99c(config)# int se0/1
Rt99c(config-if)# desc WAN link to Rt99b for backup
Rt99c(config-if)# ip addr 192.168.199.238 255.255.255.252
Rt99c(config-if)# clock rate 64000
Rt99c(config-if)# no shut

（6）配置 Rt99b 增加的接口

Rt99b(config)# int se0/2
Rt99b(config-if)# desc WAN link to Rt99a for backup
Rt99b(config-if)# ip addr 192.168.199.234 255.255.255.252
Rt99b(config-if)# no shut

Rt99b(config-if)# int se0/3
Rt99b(config-if)# desc WAN link to Rt99c for backup
Rt99b(config-if)# ip addr 192.18.199.237 255.255.255.252
Rt99b(config-if)# no shut

8．配置与检验浮动静态路由

静态路由对应的通信路径为主链路，浮动静态路由对应的通信路径为备用链路。当主链路正常时,浮动静态路由不起作用;当主链路中断时,浮动静态路由便取代静态路由起作用。

在配置时,静态路由的管理距离默认为1,浮动静态路由的管理距离应大于1。

(1) 在 Rt99a 上配置浮动静态路由

```
Rt99a(config)# ip route 0.0.0.0 0.0.0.0 192.168.199.234 6    //配置浮动静态路由
Rt99a(config)# end
Rt99a# show ip route
Gateway of last resort is 192.168.199.226 to network 0.0.0.0
        192.168.199.0/24 is variably subnetted, 3 subnets, 2 masks
C       192.168.199.0/25 is directly connected, FastEthernet0/0
C       192.168.199.224/30 is directly connected, Serial0/0
C       192.168.199.232/30 is directly connected, Serial0/1
S*      0.0.0.0/0 [1/0] via 192.168.199.226
//在主链路正常时看不到浮动静态路由
```

(2) 在 Rt99c 上配置浮动静态路由

```
Rt99c(config)# ip route 0.0.0.0 0.0.0.0 192.168.199.237 4    //配置浮动静态路由
Rt99c(config)# end
Rt99c# show ip route
Gateway of last resort is 192.168.199.229 to network 0.0.0.0
        192.168.199.0/24 is variably subnetted, 3 subnets, 2 masks
C       192.168.199.128/26 is directly connected, FastEthernet0/0
C       192.168.199.228/30 is directly connected, Serial0/0
C       192.168.199.236/30 is directly connected, Serial0/1
S*      0.0.0.0/0 [1/0] via 192.168.199.229
//在主链路正常时看不到浮动静态路由
```

(3) 在 Rt99b 上配置浮动静态路由

```
Rt99b(config)# ip route 192.168.199.0 255.255.255.128 192.168.199.233 6
Rt99b(config)# ip route 192.168.199.128 255.255.255.192 192.168.199.238 4
Rt99b(config)# end
Rt99b# show ip route
Gateway of last resort is not set
        192.18.199.0/30 is subnetted, 1 subnets
C       192.18.199.236 is directly connected, Serial0/1/1
        192.168.199.0/24 is variably subnetted, 6 subnets, 4 masks
S       192.168.199.0/25 [1/0] via 192.168.199.225
S       192.168.199.128/26 [1/0] via 192.168.199.230
C       192.168.199.192/27 is directly connected, FastEthernet0/0
C       192.168.199.224/30 is directly connected, Serial0/0
C       192.168.199.228/30 is directly connected, Serial0/1
C       192.168.199.232/30 is directly connected, Serial0/2
```

C   192.168.199.236/30 is directly connected, Serial0/3
//在主链路正常时看不到浮动静态路由

(4) 在 Rt99a 上查看配置文件

```
Rt99a# show running-config                //查看运行配置
Building configuration...
……
interface Serial0/0/0
  description WAN link to Rt99b in building B
  ip address 192.168.199.225 255.255.255.252
  clock rate 2000000
//以上 4 行显示了主链路配置
!
interface Serial0/1/0
  description WAN link to Rt99b for backup
  ip address 192.168.199.233 255.255.255.252
  clock rate 64000
//以上 4 行显示了备份链路配置
!
……
!
ip classless
ip route 0.0.0.0 0.0.0.0 192.168.199.226
ip route 0.0.0.0 0.0.0.0 192.168.199.234 6
//以上 2 行显示了静态路由配置
```

(5) 在 Rt99b 上查看配置文件

```
Rt99b# show running-config
Building configuration...
……
!
interface Serial0/0/0
  description WAN link to Rt99a in building A
  ip address 192.168.199.226 255.255.255.252
//以上 3 行显示了到 Rt99a 的主链路配置
!
interface Serial0/0/1
  description WAN link to Rt99c in building C
  ip address 192.168.199.229 255.255.255.252
//以上 3 行显示了到 Rt99c 的主链路配置
!
interface Serial0/1/0
  description WAN link to Rt99a for backup
  ip address 192.168.199.234 255.255.255.252
```

//以上 3 行显示了到 Rt99a 的备份链路配置
!
interface Serial0/1/1
  description WAN link to Rt99c for backup
  ip address 192.18.199.237 255.255.255.252
//以上 3 行显示了到 Rt99c 的备份链路配置
!
……
!
ip classless
ip route 192.168.199.0 255.255.255.128 192.168.199.225
ip route 192.168.199.128 255.255.255.192 192.168.199.230
ip route 192.168.199.0 255.255.255.128 192.168.199.233 6
ip route 192.168.199.128 255.255.255.192 192.168.199.238 4
//以上 4 行显示了静态路由配置
!

（6）在 Rt99c 上查看配置文件

Rt99c# show running-config
Building configuration...
……
!
interface Serial0/0/0
  description WAN link to Rt99b in building B
  ip address 192.168.199.230 255.255.255.252
  clock rate 2000000
//以上 4 行显示了主链路配置
!
interface Serial0/1/0
  description WAN link to Rt99b for backup
  ip address 192.168.199.238 255.255.255.252
  clock rate 64000
//以上 4 行显示了备份链路配置
!
……
!
ip classless
ip route 0.0.0.0 0.0.0.0 192.168.199.229
ip route 0.0.0.0 0.0.0.0 192.168.199.237 4
//以上 2 行显示了静态路由配置
!

（7）保存每台路由器的配置

Rt99a# write

Rt99b# write
Rt99c# write

9. 模拟主链路中断与恢复

单击【保存】按钮将相关配置保存到 P06005.pkt 文件中备用，再将其另存为文件名为 P06006.pkt 的文件，然后继续进行实训工作。

为了检验备份链路是否起作用，需要模拟主链路中断的故障，这可通过关闭相应的接口来实现。开启相应的接口即可恢复主链路。

（1）关闭接口以模拟故障

Rt99b(config)# int se0/0
Rt99b(config-if)# shutdown   //关闭到 Rt99a 的主链路接口
Rt99b(config-if)# int se0/1
Rt99b(config-if)# shutdown   //关闭到 Rt99c 的主链路接口

（2）查看 Rt99a 的路由表

Rt99a# show ip route
Gateway of last resort is 192.168.199.234 to network 0.0.0.0   //注意这一行的变动
    192.168.199.0/24 is variably subnetted, 2 subnets, 2 masks
C   192.168.199.0/25 is directly connected, FastEthernet0/0
C   192.168.199.232/30 is directly connected, Serial0/1
S*   0.0.0.0/0 [6/0] via 192.168.199.234
//在主链路中断时看到了浮动静态路由（管理距离为 6）
请注意浮动静态路由的管理距离和下一跳的 IP 地址。

（3）查看 Rt99c 的路由表

Rt99c# show ip route
Gateway of last resort is 192.168.199.237 to network 0.0.0.0   //注意这一行的变动
    192.168.199.0/24 is variably subnetted, 2 subnets, 2 masks
C   192.168.199.128/26 is directly connected, FastEthernet0/0
C   192.168.199.236/30 is directly connected, Serial0/1
S*   0.0.0.0/0 [4/0] via 192.168.199.237
//在主链路中断时看到了浮动静态路由（管理距离为 4）
请注意浮动静态路由的管理距离和下一跳的 IP 地址。

（4）查看 Rt99b 的路由表

Rt99b# show ip route
Gateway of last resort is not set
    192.168.199.0/24 is variably subnetted, 5 subnets, 4 masks
S   192.168.199.0/25 [6/0] via 192.168.199.233
S   192.168.199.128/26 [4/0] via 192.168.199.238
C   192.168.199.192/27 is directly connected, FastEthernet0/0

```
C         192.168.199.232/30 is directly connected, Serial0/2
C         192.168.199.236/30 is directly connected, Serial0/3
```
//在主链路中断时看到了浮动静态路由（管理距离分别为 6 和 4）

请注意浮动静态路由的管理距离和下一跳的 IP 地址。

（5）在 Pc99a 上测试网络连通性

```
C:\> ping -n 2 192.168.199.222
!!!!!
C:\> ping -n 2 192.168.199.190
!!!!!
```

测试结果证明，网络还是连通的。

（6）开启接口以模拟链路恢复

```
Rt99b(config)# int se0/0
Rt99b(config-if)# no shutdown        //开启到 Rt99a 的主链路接口
Rt99b(config-if)# int se0/1
Rt99b(config-if)# no shutdown        //开启到 Rt99c 的主链路接口
```

（7）查看 Rt99a 的路由表

```
Rt99a# show ip route
Gateway of last resort is 192.168.199.226 to network 0.0.0.0
     192.168.199.0/24 is variably subnetted, 3 subnets, 2 masks
C        192.168.199.0/25 is directly connected, FastEthernet0/0
C        192.168.199.224/30 is directly connected, Serial0/0
C        192.168.199.232/30 is directly connected, Serial0/1
S*       0.0.0.0/0 [1/0] via 192.168.199.226
```

请注意有变动的部分。

（8）查看 Rt99c 的路由表

```
Rt99c# show ip route
Gateway of last resort is 192.168.199.229 to network 0.0.0.0
     192.168.199.0/24 is variably subnetted, 3 subnets, 2 masks
C        192.168.199.128/26 is directly connected, FastEthernet0/0
C        192.168.199.228/30 is directly connected, Serial0/0
C        192.168.199.236/30 is directly connected, Serial0/1
```

请注意有变动的部分。

（9）查看 Rt99b 的路由表

```
Rt99b# show ip route
Gateway of last resort is not set
     192.18.199.0/30 is subnetted, 1 subnets
C        192.18.199.236 is directly connected, Serial0/1/1
```

```
         192.168.199.0/24 is variably subnetted, 6 subnets, 4 masks
S        192.168.199.0/25 [1/0] via 192.168.199.225
S        192.168.199.128/26 [1/0] via 192.168.199.230
C        192.168.199.192/27 is directly connected, FastEthernet0/0
C        192.168.199.224/30 is directly connected, Serial0/0
C        192.168.199.228/30 is directly connected, Serial0/1
C        192.168.199.232/30 is directly connected, Serial0/2
C        192.168.199.236/30 is directly connected, Serial0/3
```

请注意有变动的部分。

（10）在 PC99c 上测试网络连通性

```
C:\> ping -n 2 192.168.199.126
!!!!!
C:\> ping -n 2 192.168.199.222
!!!!!
```

从以上各测试结果可知，网络已经全部连通，而且备份链路也起作用！

请注意，当主链路中断时备份链路和浮动静态路由自动起作用；当主链路恢复时备份链路和浮动静态路由又自动转变为备用状态。

请牢记，修改配置后都要执行保存配置的命令：write 或 copy run start。

单击【保存】按钮将相关配置保存到 P06006.pkt 文件中备用，然后再把它另存为文件名为 P06007.pkt 的文件。

学习总结

通过本项目，我们不仅学会了恰当、灵活地配置静态路由，而且还学会了浮动静态路由的配置方法，熟练地掌握了路由器的串行接口和以太网接口的配置、各种信息查询，以及故障模拟和网络调试等方法，另外，我们也学习了子网地址和子网掩码的恰当应用。

本项目的重点是静态路由和默认路由的配置，特别是备份链路和浮动静态路由的配置。默认路由是静态路由一种特例。

课后作业

完成上面的配置和调试，将实训过程的截图按顺序粘贴到一个 Word 文件里并用适当的文字说明你对它的理解；总结本次实训所需要的主要命令及其作用，作为实训报告上交。

根据项目 01 中的路由器命名规则，务必把路由器命名为 RtIDa，其中 ID 为各自学号的最后两位数字。实训报告一律以"ID 姓名项目号.doc"命名，网络拓扑及其配置也以"ID 姓名项目号.pkt"为文件名保存并上交。例如，张三的 ID 为 03，他的文件名为"03 张三 06.doc"和"03 张三 06.pkt"。

项目 06　静态路由综合配置

思考题

　　随着网络中路由器数量的增加，静态路由的配置将迅速变得复杂和烦琐。因此，大规模的网络不适合采用静态路由，网络拓扑变化频繁的网络也不适合采用静态路由，而应配置动态路由。那么，怎样配置动态路由呢？
　　在下一个项目中我们将学习有关的知识和技术。

项目 07 PPP 与 RIP 配置

项目描述

若你是某单位的网络管理员,你需要构建连接两个园区的网络。每个园区网是一个以太网,两个园区网之间距离较远,由专线连接构成一个广域网。为了通信安全,你决定广域网连接采用 PPP 封装和 CHAP 认证。由于网络规模小,你决定采用动态路由。

网络拓扑

网络拓扑如图 7-1 所示。

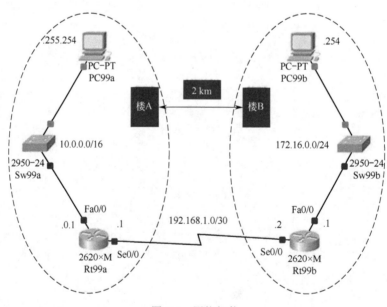

图 7-1 网络拓扑

学习目标

- 重点掌握串行接口和 PPP 的配置方法;

项目 07　PPP 与 RIP 配置

- 重点掌握 RIPv2 的配置方法；
- 掌握查看路由信息的方法；
- 掌握验证配置是否正确的方法；
- 掌握查看所有配置信息的方法。

实训任务分解

① 配置路由器。
② 查看和配置串行接口。
③ 配置以太网接口。
④ 配置 PC 的 IP 地址。
⑤ 完成网络连通性测试。
⑥ 保存配置。
⑦ 配置 RIPv2。

知识点介绍

① 随着网络中路由器数量的增加，静态路由的配置将迅速变得复杂和烦琐。因此，大规模的网络不适合采用静态路由，网络拓扑变化频繁的网络也不适合采用静态路由，而应配置动态路由。

② 动态路由是由网络中的路由器按路由协议交换路由信息而自动产生的。动态路由协议有 RIP、IGRP、EIGRP、OSPF 等。

③ RIP（Routing Information Protocol，路由信息协议）有 RIPv1 和 RIPv2 两个版本，本项目我们将重点学习 RIPv2 的配置和串行接口的 PPP 封装及其 CHAP 认证配置。

实训过程

首先，构建如图 7-1 所示的网络，注意两台路由器都要增加一块 WIC-1T 模块，这样才能用串行线路相连接。

1．配置路由器

（1）配置 Rt99a

```
Router(config)# host Rt99a
Rt99a(config)# enable secret 99secret
Rt99a(config)# line con 0
```

```
Rt99a(config-line)# logg sync
Rt99a(config-line)# line vty 0 15
Rt99a(config-line)# password 99vty015
Rt99a(config-line)# login
Rt99a(config-line)# logg sync
Rt99a(config-line)# exec-timeout 0 0
Rt99a(config-line)# exit
Rt99a(config)# service password-encr
Rt99a(config)# no ip domain-lookup
Rt99a(config)# end
```

（2）配置 Rt99b

```
Router(config)# host Rt99b
Rt99b(config)# enable secret 99secret
Rt99b(config)# line con 0
Rt99b(config-line)# logg sync
Rt99b(config-line)# exec-timeout 0 0
Rt99b(config-line)# line vty 0 15
Rt99b(config-line)# password 99vty015
Rt99b(config-line)# login
Rt99b(config-line)# logg sync
Rt99b(config-line)# exec-timeout 0 0
Rt99b(config-line)# exit
Rt99b(config)# service password-encr
Rt99b(config)# no ip domain-lookup
Rt99b(config)# end
```

2. 查看和配置串行接口

（1）查看 Rt99a 串行接口

```
Rt99a# show controller se0/0
Interface Serial0/0
Hardware is PowerQUICC MPC860
DCE V.35, no clock                                    //DCE 端
idb at 0x81081AC4, driver data structure at 0x81084AC0
SCC Registers:
…..
--More--
```

（2）查看 Rt99b 串行接口

```
Rt99b# show controller se0/0
Interface Serial0/0
```

```
Hardware is PowerQUICC MPC860
DTE V.35 clocks stopped.                              //DTE 端
idb at 0x81081AC4, driver data structure at 0x81084AC0
SCC Registers:
…..
--More—
```

（3）配置 Rt99a 串行接口

```
Rt99a(config)# int se0/0
Rt99a(config-if)# desc WAN link to Rt99b
Rt99a(config-if)# ip addr 192.168.1.1 255.255.255.252
Rt99a(config-if)# clock rate 128000
Rt99a(config-if)# encapsulation ppp              //配置封装协议为 PPP
Rt99a(config-if)# ppp authentication chap        //配置 PPP 认证方式为 CHAP
Rt99a(config-if)# no shut
Rt99a(config-if)# exit
Rt99a(config)# username Rt99b password 99Hello   //配置用户名和密码
Rt99a(config)# end
```

注意：用户名为对方路由器名，双方密码要一致。

（4）配置 Rt99b 串行接口

```
Rt99b(config)# username Rt99a password 99Hello   //配置用户名和密码
Rt99b(config)# int se0/0
Rt99b(config-if)# desc WAN Link to Rt99a
Rt99b(config-if)# ip addr 192.168.1.2 255.255.255.252
Rt99b(config-if)# encapsulation ppp              //配置封装协议为 PPP
Rt99b(config-if)# ppp authentication chap        //配置 PPP 认证方式为 CHAP
Rt99b(config-if)# no shut
Rt99b(config-if)# end
```

注意：用户名为对方路由器名，双方密码要一致。

（5）在 Rt99a 上测试 WAN 网络连通性

```
Rt99a# ping 192.168.1.2
!!!!!
```

（6）在 Rt99b 上测试 WAN 网络连通性

```
Rt99b# ping 192.168.1.1
!!!!!
```

3. 配置以太网接口

（1）配置 Rt99a 的以太网接口

```
Rt99a(config)# int fa0/0
Rt99a(config-if)# desc LAN Link to Sw99a
```

```
Rt99a(config-if)# ip addr 10.0.0.1 255.255.0.0
Rt99a(config-if)# no shut
Rt99a(config-if)# end
```

(2) 配置 Rt99b 的以太网接口

```
Rt99b(config)# int fa0/0
Rt99b(config-if)# desc LAN Link to Sw99b
Rt99b(config-if)# ip addr 172.16.0.1 255.255.255.0
Rt99b(config-if)# no shut
Rt99b(config-if)# end
```

4．配置 PC 的 IP 地址

在各台 PC 上按如下数据配置 IP 地址、子网掩码和默认网关。

① PC99a 的配置。
- IP 地址：10.0.255.254；
- 子网掩码：255.255.0.0；
- 默认网关：10.0.0.1。

② PC99b 的配置。
- IP 地址：172.16.0.254；
- 子网掩码：255.255.255.0；
- 默认网关：172.16.0.1。

5．完成网络连通性测试

(1) 在 Rt99a 上测试网络连通性

```
Rt99a# ping 192.168.1.2                         //通
!!!!!
Rt99a# ping 10.0.255.254                        //通
.!!!!
Rt99a# ping 172.16.0.254                        //不通
......
```

(2) 在 Rt99b 上测试网络连通性

```
Rt99b# ping 192.168.1.1                         //通
!!!!!
Rt99b# ping 172.16.0.254                        //通
.!!!!
Rt99b# ping 10.0.255.254                        //不通
......
```

6．保存配置

```
Rt99a# copy run start
Rt99b# copy run start
```

项目 07　PPP 与 RIP 配置

注意：为了避免后面配置其他动态路由时重复前面的配置，请在这里执行上面的命令保存配置，先将执行结果保存为文件名为 P07001.pkt 的文件备用，然后再将其另存为文件名为 P07002.pkt 的文件，这样下面的配置可在此文件的基础上开始，否则一旦退出就只能从头开始了。

7. 配置 RIPv2

前面的测试表明，路由器直连的网络都已连通，而非直连的网络间还不通。这时，只要配置好路由便都可连通了。这里要学习配置 RIPv2。

（1）查看 Rt99a 的路由信息

```
Rt99a# show ip route
Gateway of last resort is not set
10.0.0.0/16 is subnetted, 1 subnets
C   10.0.0.0 is directly connected, FastEthernet0/0
192.168.1.0/24 is variably subnetted, 2 subnets, 2 masks
C   192.168.1.0/30 is directly connected, Serial0/0
C   192.168.1.2/32 is directly connected, Serial0/0
//此时只有 3 条直连路由
```

（2）查看 Rt99b 的路由信息

```
Rt99b#show ip route
Gateway of last resort is not set
172.16.0.0/24 is subnetted, 1 subnets
C   172.16.0.0 is directly connected, FastEthernet0/0
192.168.1.0/24 is variably subnetted, 2 subnets, 2 masks
C   192.168.1.0/30 is directly connected, Serial0/0
C   192.168.1.1/32 is directly connected, Serial0/0
//此时只有 3 条直连路由
```

（3）在 Rt99a 上配置 RIPv2

```
Rt99a(config)# router rip                          //指定路由协议
Rt99a(config-router)# version 2                    //指定版本 2
Rt99a(config-router)# no auto-summary              //取消自动汇总功能
Rt99a(config-router)# network 192.168.1.0          //配置通告网络
Rt99a(config-router)# network 10.0.0.0             //配置通告网络
Rt99a(config-router)# passive-interface fa0/0      //禁止在 Fa0/0 上广播路由信息
Rt99a(config-router)# end
```

（4）在 Rt99b 上配置 RIPv2

```
Rt99b(config)# router rip
Rt99b(config-router)# version 2                    //配置 RIPv2
Rt99b(config-router)# no auto-summary              //取消自动汇总功能
```

```
Rt99b(config-router)# network 192.168.1.0          //配置通告网络
Rt99b(config-router)# network 172.16.0.0           //配置通告网络
Rt99b(config-router)# passive-interface fa0/0      //禁止在 Fa0/0 上广播路由信息
Rt99b(config-router)# end
```

（5）查看 Rt99a 的路由信息

```
Rt99a# show ip route
Gateway of last resort is not set
10.0.0.0/16 is subnetted, 1 subnets
C  10.0.0.0 is directly connected, FastEthernet0/0
172.16.0.0/24 is subnetted, 1 subnets
R  172.16.0.0 [120/1] via 192.168.1.2, 00:00:12, Serial0/0    //动态路由
192.168.1.0/24 is variably subnetted, 2 subnets, 2 masks
C  192.168.1.0/30 is directly connected, Serial0/0
C  192.168.1.2/32 is directly connected, Serial0/0
//此时已多了动态路由。只有双方同时运行 RIPv2 才能看到
```

（6）查看 Rt99b 的路由信息

```
Rt99b# show ip route
Gateway of last resort is not set
10.0.0.0/16 is subnetted, 1 subnets
R  10.0.0.0 [120/1] via 192.168.1.1, 00:00:23, Serial0/0     //动态路由
172.16.0.0/24 is subnetted, 1 subnets
C  172.16.0.0 is directly connected, FastEthernet0/0
192.168.1.0/24 is variably subnetted, 2 subnets, 2 masks
C  192.168.1.0/30 is directly connected, Serial0/0
C  192.168.1.1/32 is directly connected, Serial0/0
//此时已多了动态路由。只有双方同时运行 RIPv2 才能看到
```

（7）在 Rt99a 上测试网络连通性

```
Rt99a# ping 172.16.0.254                           //通
!!!!!
```

（8）在 Rt99b 上测试网络连通性

```
Rt99b# ping 10.0.255.254                           //通
!!!!!
```

（9）在 Rt99a 上查看路由协议

```
Rt99a# show ip protocols
Routing Protocol is "rip"
……
```

（10）在 Rt99b 上查看路由协议

```
Rt99b# show ip protocols
```

项目 07 PPP 与 RIP 配置

```
Routing Protocol is "rip"
……
```

（11）查看 Rt99a 的配置

```
Rt99a# show run                    //显示所有配置信息
Building configuration...
Current configuration : 1403 bytes
……
interface FastEthernet0/0
 description LAN Link to Sw99a
 ip address 10.0.0.1 255.255.0.0
 duplex auto
 speed auto
!
interface Serial0/0
 description WAN link to Rt99b
 ip address 192.168.1.1 255.255.255.252
 encapsulation ppp
 ppp authentication chap
 clock rate 128000
!
router rip
 version 2
 passive-interface FastEthernet0/0
 network 10.0.0.0
 network 192.168.1.0
 no auto-summary
//以上 6 行显示了 Rt99a 上的路由协议配置情况
!
……
```

（12）查看 Rt99b 的配置

在 Rt99b 上看到的配置与上面的相似，故省略。

（13）保存配置

```
Rt99a# write                       //保存配置
Rt99b# write                       //保存配置
```

保存配置备用，然后再将结果另存为文件名为 P07003.pkt 的文件。

学习总结

串行链路上常用的两种封装协议为 HDLC 和 PPP，前者只在 Cisco 设备间使用，后者可以用于不同厂商的设备间。PPP 比 HDLC 有较多的功能。

PPP 有两种认证方式：CHAP 和 PAP，CHAP 比 PAP 具有较好的安全性能。在配置 PPP 认证时要求用户名为对方路由器名，双方密码必须一致。

动态路由协议包括距离向量路由协议和链路状态路由协议。RIP 曾是使用最广泛的距离向量路由协议，是为小型网络环境设计的，因为其路由学习及路由更新会产生较大的流量，所以会占用过多的带宽。

为了避免路由环路，RIP 采用水平分割、毒性逆转、定义最大跳数、闪式更新、抑制计时五个机制。RIP 的管理距离默认是 120。RIP 有 2 个版本：版本 1 和版本 2。版本 1 已基本被淘汰。RIPv1 和 RIPv2 的主要区别如表 7-1 所示。

表 7-1　RIPv1 和 RIPv2 的主要区别

RIPv1	RIPv2
在路由更新信息中不携带子网信息	在路由更新信息中携带子网信息
不支持 VLSM 和 CIDR	支持 VLSM 和 CIDR
有类别（Classful）路由协议	无类别（Classless）路由协议
不提供认证	提供明文和 MD5 认证
采用广播方式更新	采用组播（224.0.0.9）方式更新

课后作业

完成上面的模拟实训，将实训过程的截图按顺序粘贴到一个 Word 文件里并用适当的文字说明你对它的理解；总结本次实训所需要的主要命令及其作用，作为实训报告上交。

根据项目 01 中的路由器命名规则，务必把路由器命名为 RtIDa，其中 ID 为各自学号的最后两位数字。实训报告一律以"ID 姓名项目号.doc"为文件名命名，网络拓扑及其配置也以"ID 姓名项目号.pkt"为文件名保存并上交。例如，张三的 ID 为 03，他的文件名为"03 张三 07.doc"和"03 张三 07.pkt"。

思考题

若将 RIP 改为版本 1，结果将有何不同？（对这个问题感兴趣的同学可以自己试一下）。
除了 RIP，还有哪些动态路由协议？换成别的协议又将如何配置？
在下一个项目中我们将学习解决办法。

项目 08　EIGRP 与 OSPF 配置

项目描述

若你是某单位的网络管理员，你需要掌握静态路由和常用动态路由的配置方法，并根据网络设备、网络结构和规模适当地选择和配置静态路由或动态路由。

网络拓扑

网络拓扑如图 8-1 所示。

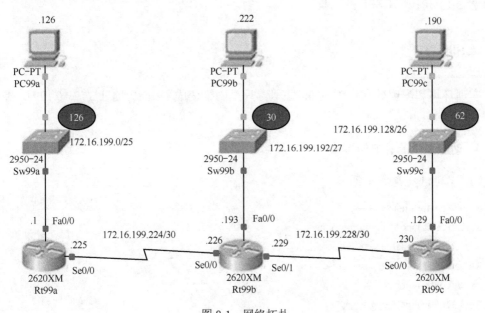

图 8-1　网络拓扑

学习目标

- 重点掌握 EIGRP 的配置方法；
- 重点掌握 OSPF 的配置方法；

- 能利用 Loopback 接口设置路由器的 ID；
- 掌握查看路由的方法；
- 掌握调试网络的方法。

实训任务分解

① 动态路由配置前的准备。
② EIGRP 的配置与检验。
③ OSPF 的配置与检验。

知识点介绍

动态路由协议分内部网关协议（IGP）和外部网关协议（EGP），我们只学习 IGP。

RIP、IGRP、EIGRP、OSPF 都属于 IGP 动态路由协议。就如 RIPv2 是 RIPv1 的改进版一样，EIGRP 也是 IGRP 的改进版。因此，我们只学习 RIPv2 和 EIGRP 的配置。本项目我们将学习 EIGRP 和 OSPF 的配置。

实训过程

实训环境和配置数据与项目 06 相似，只是网络变成了由 IP 地址 172.16.199.0 划分的子网，如图 8-1 所示。

1. 动态路由配置前的准备

（1）完成 Rt99a 基本配置

```
Router(config)#host Rt99a
Rt99a(config)#no ip domain-lookup
Rt99a(config)#enable secret 99secret
Rt99a(config)#service password-encr
Rt99a(config)#line con 0
Rt99a(config-line)#logg sync
Rt99a(config-line)#exec-timeout 0 0
Rt99a(config-line)#exit
```

（2）完成 Rt99b 基本配置

```
Router(config)#host Rt99b
Rt99b(config)#no ip domain-lookup
Rt99b(config)#service password-encr
```

项目 08　EIGRP 与 OSPF 配置

Rt99b(config)#line con 0
Rt99b(config-line)#logg sync
Rt99b(config-line)#exec-timeout 0 0
Rt99b(config-line)#exit

（3）完成 Rt99c 基本配置

Router(config)#host Rt99c
Rt99c(config)#no ip domain-lookup
Rt99c(config)#service password-encr
Rt99c(config)#enable secret 99secret
Rt99c(config)#line con 0
Rt99c(config-line)#logg sync
Rt99c(config-line)#exec-timeout 0 0
Rt99c(config-line)#exit

注意：各路由器都没有开启远程登录功能。

（4）配置 Rt99a 的接口

Rt99a(config)#int fa0/0
Rt99a(config-if)#desc LAN link to Sw99a in building A
Rt99a(config-if)#ip addr 172.16.199.1 255.255.255.128
Rt99a(config-if)#no shut
Rt99a(config-if)#int se0/0
Rt99a(config-if)#desc WAN link to Rt99b in building B
Rt99a(config-if)#ip addr 172.16.199.225 255.255.255.252
Rt99a(config-if)#clock rate 2000000
Rt99a(config-if)#no shut

（5）配置 Rt99c 的接口

Rt99c(config)#int fa0/0
Rt99c(config-if)#desc LAN link to Sw99c in building C
Rt99c(config-if)#ip addr 172.16.199.129 255.255.255.192
Rt99c(config-if)#no shut
Rt99c(config-if)#int se0/0
Rt99c(config-if)#desc WAN link to Rt99b in building B
Rt99c(config-if)#ip addr 172.16.199.230 255.255.255.252
Rt99c(config-if)#clock rate 1000000
Rt99c(config-if)#no shut

（6）配置 Rt99b 的接口

Rt99b(config)#int fa0/0
Rt99b(config-if)#desc LAN link to Sw99b in building B
Rt99b(config-if)#ip addr 172.16.199.193 255.255.255.224
Rt99b(config-if)#no shut

```
Rt99b(config-if)#int se0/0
Rt99b(config-if)#desc WAN link to Rt99a in building A
Rt99b(config-if)#ip addr 172.16.199.226 255.255.255.252
Rt99b(config-if)#no shut
Rt99b(config-if)#int se0/1
Rt99b(config-if)#desc WAN link to Rt99c in building C
Rt99b(config-if)#ip addr 172.16.199.229 255.255.255.252
Rt99b(config-if)#no shut
```

（7）配置 PC 的 IP 地址

在各台 PC 上按如下数据配置 IP 地址、子网掩码和默认网关：
① PC99a 的配置。
- IP 地址：172.16.199.126；
- 子网掩码：255.255.255.128；
- 默认网关：172.16.199.1。

② PC99b 的配置。
- IP 地址：172.16.199.222；
- 子网掩码：255.255.255.224；
- 默认网关：172.16.199.193。

③ PC99c 的配置。
- IP 地址：172.16.199.190；
- 子网掩码：255.255.255.192；
- 默认网关：172.16.199.129。

（8）在 Rt99a 上测试网络连通性

```
Rt99a#ping 172.16.199.126        //通
.!!!!
Rt99a#ping 172.16.199.226        //通
!!!!!
```

（9）在 Rt99b 上测试网络连通性

```
Rt99b#ping 172.16.199.222        //通
.!!!!
Rt99b#ping 172.16.199.225        //通
!!!!!
Rt99b#ping 172.16.199.230        //通
!!!!!
```

（10）在 Rt99c 上测试网络连通性

```
Rt99c#ping 172.16.199.190        //通
.!!!!
```

项目 08　EIGRP 与 OSPF 配置

```
Rt99c#ping 172.16.199.229          //通
!!!!!
```

（11）保存配置

```
Rt99a# write
Rt99b# write
Rt99c# write
```

单击【保存】按钮，先将相关配置保存为文件名为 P08001.pkt 的文件备用，再将其另存为文件名为 P08002.pkt 的文件，然后继续进行实训工作。

2. EIGRP 的配置与检验

完成了上面的准备工作后，就可开始配置动态路由协议了。

注意：运行 EIGRP 的整个网络的 AS 号必须一致，其范围为 1～65535，如 333。

（1）查看 Rt99a 的路由表

```
Rt99a# show ip route
Gateway of last resort is not set
172.16.0.0/16 is variably subnetted, 2 subnets, 2 masks
C 172.16.199.0/25 is directly connected, FastEthernet0/0
C 172.16.199.224/30 is directly connected, Serial0/0
//共有 5 个子网，但只有 2 条直连路由
```

（2）查看 Rt99b 的路由表

```
Rt99b# show ip route
Gateway of last resort is not set
172.16.0.0/16 is variably subnetted, 3 subnets, 2 masks
C 172.16.199.192/27 is directly connected, FastEthernet0/0
C 172.16.199.224/30 is directly connected, Serial0/0
C 172.16.199.228/30 is directly connected, Serial0/1
//共有 5 个子网，但只有 3 条直连路由
```

（3）查看 Rt99c 的路由表

```
Rt99c# show ip route
Gateway of last resort is not set
172.16.0.0/16 is variably subnetted, 2 subnets, 2 masks
C 172.16.199.128/26 is directly connected, FastEthernet0/0
C 172.16.199.228/30 is directly connected, Serial0/0
//共有 5 个子网，但只有 2 条直连路由
```

（4）在 Rt99a 上配置 EIGRP

```
Rt99a(config)# router eigrp 333                //启动 EIGRP 进程
Rt99a(config-router)# no auto-summary          //取消自动汇总功能
```

```
Rt99a(config-router)# network 172.16.0.0           //通告网络
Rt99a(config-router)# passive-interface fa0/0      //被动接口
Rt99a(config-router)# end
```

（5）在 Rt99b 上配置 EIGRP

```
Rt99b(config)# router eigrp 333                    //启动 EIGRP 进程
Rt99b(config-router)# no auto-summary              //取消自动汇总功能
Rt99b(config-router)# network 172.16.0.0           //通告网络
Rt99b(config-router)# passive-interface fa0/0      //被动接口
Rt99b(config-router)# end
```

（6）在 Rt99c 上配置 EIGRP

```
Rt99c(config)# router eigrp 333                    //启动 EIGRP 进程
Rt99c(config-router)# no auto-summary              //取消自动汇总功能
Rt99c(config-router)# network 172.16.0.0           //通告网络
Rt99c(config-router)# passive-interface fa0/0      //被动接口
Rt99c(config-router)# end
```

（7）在 Rt99a 上查看路由信息

```
Rt99a# show ip route
Gateway of last resort is not set
172.16.0.0/16 is variably subnetted, 5 subnets, 4 masks
C    172.16.199.0/25 is directly connected, FastEthernet0/0
D    172.16.199.128/26 [90/2684416] via 172.16.199.226, 00:03:35, Serial0/0
D    172.16.199.192/27 [90/2172416] via 172.16.199.226, 00:06:03, Serial0/0
C    172.16.199.224/30 is directly connected, Serial0/0
D    172.16.199.228/30 [90/2681856] via 172.16.199.226, 00:06:03, Serial0/0
//以 D 开头的 3 行所示为动态路由
```

（8）在 Rt99b 上查看路由信息

```
Rt99b# show ip route
Gateway of last resort is not set
172.16.0.0/16 is variably subnetted, 5 subnets, 4 masks
D    172.16.199.0/25 [90/2172416] via 172.16.199.225, 00:07:19, Serial0/0
D    172.16.199.128/26 [90/2172416] via 172.16.199.230, 00:04:51, Serial0/1
C    172.16.199.192/27 is directly connected, FastEthernet0/0
C    172.16.199.224/30 is directly connected, Serial0/0
C    172.16.199.228/30 is directly connected, Serial0/1
//以 D 开头的 2 行所示为动态路由
```

（9）在 Rt99c 上查看路由信息

```
Rt99c# show ip route
Gateway of last resort is not set
172.16.0.0/16 is variably subnetted, 5 subnets, 4 masks
```

项目 08　EIGRP 与 OSPF 配置

```
D 172.16.199.0/25 [90/2172416] via 172.16.199.229, 00:05:43, Serial0/0
C 172.16.199.128/26 is directly connected, FastEthernet0/0
D 172.16.199.192/27 [90/2172416] via 172.16.199.229, 00:05:43, Serial0/0
D 172.16.199.224/30 [90/2681856] via 172.16.199.229, 00:05:43, Serial0/0
C 172.16.199.228/30 is directly connected, Serial0/0
//以 D 开头的 3 行所示为动态路由
```

与配置 EIGRP 前对比，发现各路由器的路由信息中都多了以 D 开头的行。

（10）在 Rt99a 上测试网络连通性

```
Rt99a# ping 172.16.199.222              //通
!!!!!
Rt99a# ping 172.16.199.190              //通
!!!!!
```

（11）在 Rt99b 上测试网络连通性

```
Rt99b#ping 172.16.199.126               //通
!!!!!
Rt99b#ping 172.16.199.190               //通
!!!!!
```

（12）在 Rt99c 上测试网络连通性

```
Rt99c#ping 172.16.199.126               //通
!!!!!
Rt99c#ping 172.16.199.222               //通
!!!!!
```

（13）在 Rt99a 上查看路由协议相关信息

```
Rt99a# show ip protocols
Routing Protocol is "eigrp 333 "
Outgoing update filter list for all interfaces is not set
Incoming update filter list for all interfaces is not set
……
```

（14）在 Rt99b 上查看路由协议相关信息

```
Rt99b# show ip protocols
Routing Protocol is "eigrp 333 "
Outgoing update filter list for all interfaces is not set
Incoming update filter list for all interfaces is not set
……
```

（15）在 Rt99c 上查看路由协议相关信息

```
Rt99c# show ip protocols
Routing Protocol is "eigrp 333 "
```

```
Outgoing update filter list for all interfaces is not set
Incoming update filter list for all interfaces is not set
……
```

（16）保存配置

```
Rt99a# write
Rt99b# write
Rt99c# write
```

单击【保存】按钮先将相关配置保存为文件名为 P08002.pkt 的文件备用，再将其另存为文件名为 P08003.pkt 的文件，然后继续进行实训工作。

3. OSPF 的配置与检验

为了使 OSPF 易于识别路由器而使网络管理与故障排除更加容易，利用 Loopback 接口设置路由器的 ID。

注意：OSPF 的进程号可以各不相同；反掩码由 255.255.255.255 减去子网掩码获得。

（1）配置 Rt99a 的 ID

```
Rt99a(config)# int loopback0                          //指定环回接口
Rt99a(config-if)# ip addr 10.199.1.1 255.255.255.255  //配置 IP 地址
Rt99a(config-if)# exit
```

注意：环回接口总是处于 up 状态，无须用 no shutdown 命令来开启。

（2）配置 Rt99b 的 ID

```
Rt99b(config)# int loopback0                          //指定环回接口
Rt99b(config-if)# ip addr 10.199.2.2 255.255.255.255  //配置 IP 地址
Rt99b(config-if)# exit
```

注意：子网掩码为 255.255.255.255

（3）配置 Rt99c 的 ID

```
Rt99c(config)# int loopback0                          //指定环回接口
Rt99c(config-if)# ip addr 10.199.3.3 255.255.255.255  //配置 IP 地址
Rt99c(config-if)#exit
```

（4）在 Rt99a 上配置 OSPF

```
Rt99a(config)# router ospf 100                        //启动 OSPF 进程
Rt99a(config-router)# network 172.16.199.0 0.0.0.127 area 0
//0.0.0.127 为反掩码，由 255.255.255.255-255.255.255.128 获得
Rt99a(config-router)# network 172.16.199.224 0.0.0.3 area 0
//反掩码 0.0.0.3=255.255.255.255-255.255.255.252
Rt99a(config-router)# end
```

项目 08　EIGRP 与 OSPF 配置

（5）在 Rt99b 上配置 OSPF

Rt99b(config)# router ospf 200　　　　　　　　　　//启动 OSPF 进程
Rt99b(config-router)# network 172.16.199.192 0.0.0.31 area 0
Rt99b(config-router)# network 172.16.199.224 0.0.0.3 area 0
Rt99b(config-router)# network 172.16.199.228 0.0.0.3 area 0
Rt99b(config-router)# end

（6）在 Rt99c 上配置 OSPF

Rt99c(config)# router ospf 300　　　　　　　　　　//启动 OSPF 进程
Rt99c(config-router)# network 172.16.199.128 0.0.0.63 area 0
Rt99c(config-router)# network 172.16.199.228 0.0.0.3 area 0
Rt99c(config-router)# end

（7）在 Rt99a 上查看路由信息

Rt99a# show ip route
Gateway of last resort is not set
10.0.0.0/32 is subnetted, 1 subnets
C　10.199.1.1 is directly connected, Loopback0
172.16.0.0/16 is variably subnetted, 5 subnets, 4 masks
C　172.16.199.0/25 is directly connected, FastEthernet0/0
D　172.16.199.128/26 [90/2684416] via 172.16.199.226, 00:13:05, Serial0/0
D　172.16.199.192/27 [90/2172416] via 172.16.199.226, 00:13:09, Serial0/0
C　172.16.199.224/30 is directly connected, Serial0/0
D　172.16.199.228/30 [90/2681856] via 172.16.199.226, 00:13:05, Serial0/0
//起作用的仍然是采用 EIGRP 交换路由信息建立的动态路由

（8）在 Rt99b 上查看路由信息

Rt99b# show ip route
Gateway of last resort is not set
10.0.0.0/32 is subnetted, 1 subnets
C　10.199.2.2 is directly connected, Loopback0
172.16.0.0/16 is variably subnetted, 5 subnets, 4 masks
D　172.16.199.0/25 [90/2172416] via 172.16.199.225, 00:14:10, Serial0/0
D　172.16.199.128/26 [90/2172416] via 172.16.199.230, 00:14:06, Serial0/1
C　172.16.199.192/27 is directly connected, FastEthernet0/0
C　172.16.199.224/30 is directly connected, Serial0/0
C　172.16.199.228/30 is directly connected, Serial0/1
//起作用的仍然是采用 EIGRP 交换路由信息建立的动态路由

（9）查看 Rt99c 路由信息

Rt99c# show ip route
Gateway of last resort is not set
10.0.0.0/32 is subnetted, 1 subnets

C 10.199.3.3 is directly connected, Loopback0
172.16.0.0/16 is variably subnetted, 5 subnets, 4 masks
D 172.16.199.0/25 [90/2684416] via 172.16.199.229, 00:14:59, Serial0/0
C 172.16.199.128/26 is directly connected, FastEthernet0/0
D 172.16.199.192/27 [90/2172416] via 172.16.199.229, 00:14:59, Serial0/0
D 172.16.199.224/30 [90/2681856] via 172.16.199.229, 00:14:59, Serial0/0
C 172.16.199.228/30 is directly connected, Serial0/0
//起作用的仍然是采用 EIGRP 交换路由信息建立的动态路由

此时 EIGRP 和 OSPF 同时运行，因为它们的管理距离分别是 90 和 110，所以 EIGRP 优先起作用。这就是上面查看不到采用 OSPF 交换路由信息建立的动态路由的原因。

为了使 OSPF 起作用，必须停止运行 EIGRP。然后检查路由表，并与之前做比较。

（10）在 Rt99a 上停止运行 EIGRP

Rt99a(config)# no router eigrp 333　　　　　　　　　　　　//停止运行 EIGRP

（11）在 Rt99b 上停止运行 EIGRP

Rt99b(config)# no router eigrp 333　　　　　　　　　　　　//停止运行 EIGRP

（12）在 Rt99c 上停止运行 EIGRP

Rt99c(config)# no router eigrp 333　　　　　　　　　　　　//停止运行 EIGRP

（13）在 Rt99a 上查看路由信息

Rt99a# show ip route
Gateway of last resort is not set
10.0.0.0/32 is subnetted, 1 subnets
C 10.199.1.1 is directly connected, Loopback0
172.16.0.0/16 is variably subnetted, 5 subnets, 4 masks
C 172.16.199.0/25 is directly connected, FastEthernet0/0
O 172.16.199.128/26 [110/129] via 172.16.199.226, 00:08:35, Serial0/0
O 172.16.199.192/27 [110/65] via 172.16.199.226, 00:11:21, Serial0/0
C 172.16.199.224/30 is directly connected, Serial0/0
O 172.16.199.228/30 [110/128] via 172.16.199.226, 00:11:05, Serial0/0
//以 O 开头的 3 行所示为采用 OSPF 交换路由信息建立的动态路由

（14）在 Rt99b 上查看路由信息

Rt99b# show ip route
Gateway of last resort is not set
10.0.0.0/32 is subnetted, 1 subnets
C 10.199.2.2 is directly connected, Loopback0
172.16.0.0/16 is variably subnetted, 5 subnets, 4 masks
O 172.16.199.0/25 [110/65] via 172.16.199.225, 00:12:26, Serial0/0
O 172.16.199.128/26 [110/65] via 172.16.199.230, 00:09:50, Serial0/1
C 172.16.199.192/27 is directly connected, FastEthernet0/0

```
C 172.16.199.224/30 is directly connected, Serial0/0
C 172.16.199.228/30 is directly connected, Serial0/1
//以 O 开头的 2 行所示为采用 OSPF 交换路由信息建立的动态路由
```

(15) 在 Rt99c 上查看路由信息

```
Rt99c# show ip route
Gateway of last resort is not set
10.0.0.0/32 is subnetted, 1 subnets
C 10.199.3.3 is directly connected, Loopback0
172.16.0.0/16 is variably subnetted, 5 subnets, 4 masks
O 172.16.199.0/25 [110/129] via 172.16.199.229, 00:10:34, Serial0/0
C 172.16.199.128/26 is directly connected, FastEthernet0/0
O 172.16.199.192/27 [110/65] via 172.16.199.229, 00:10:34, Serial0/0
O 172.16.199.224/30 [110/128] via 172.16.199.229, 00:10:34, Serial0/0
C 172.16.199.228/30 is directly connected, Serial0/0
//以 O 开头的 3 行所示为采用 OSPF 交换路由信息建立的动态路由
```

(16) 在 Rt99a 上测试网络连通性

```
Rt99a# ping 172.16.199.222                //通
.!!!!
Rt99a# ping 172.16.199.190                //通
.!!!!
```

(17) 在 Rt99a 上查看路由协议相关信息

```
Rt99a# show ip protocols
Routing Protocol is "ospf 100"
……
Routing for Networks:                     //本路由器通告的直连网络
172.16.199.0 0.0.0.127 area 0
172.16.199.224 0.0.0.3 area 0
Routing Information Sources:              //路由信息来源
Gateway Distance Last Update
10.199.1.1 110 00:17:58
10.199.2.2 110 00:15:22
10.199.3.3 110 00:15:22
Distance: (default is 110)
```

(18) 在 Rt99b 上查看路由协议相关信息

```
Rt99b# show ip protocols
Routing Protocol is "ospf 200"
……
Routing for Networks:                     //本路由器通告的直连网络
172.16.199.192 0.0.0.31 area 0
```

172.16.199.224 0.0.0.3 area 0
172.16.199.228 0.0.0.3 area 0
Routing Information Sources: //路由信息来源
Gateway   Distance   Last Update
10.199.1.1   110   00:20:59
10.199.2.2   110   00:18:23
10.199.3.3   110   00:18:23
Distance: (default is 110)

（19）在 Rt99c 上查看路由协议相关信息

Rt99c# show ip protocols
Routing Protocol is "ospf 300"
……
Routing for Networks: //本路由器通告的直连网络
172.16.199.128 0.0.0.63 area 0
172.16.199.228 0.0.0.3 area 0
Routing Information Sources: //路由信息来源
Gateway   Distance   Last Update
10.199.1.1   110   00:21:44
10.199.2.2   110   00:19:08
10.199.3.3   110   00:19:08
Distance: (default is 110)

（20）保存配置

Rt99a# write
Rt99b# write
Rt99c# write

单击【保存】按钮先将相关配置保存为文件名为 P08003.pkt 的文件备用，再把其另存为文件名为 P08004.pkt 的文件。

学习总结

EIGRP（Enhanced Interior Gateway Routing Protocol，增强型内部网关路由协议）是 Cisco 公司开发的一个平衡混合型路由协议，它融合了距离向量和链路状态两种路由协议的优点。

反掩码是用广播地址（255.255.255.255）减去子网掩码所得到的。例如，掩码是 255.255.248.0，则反掩码是 0.0.7.255。

运行 EIGRP 的整个网络的 AS 号必须一致，其范围为 1~65535。

OSPF（Open Shortest Path First，开放最短路径优先）路由协议是典型的链路状态路由协议，具有很多优点，但配置也较复杂。

OSPF 进程 ID 的范围为 1~65535，而且只有本地含义，不同路由器的路由进程 ID 可以不同。

项目 08　EIGRP 与 OSPF 配置

为了使 OSPF 易于识别路由器而使网络管理和故障排除更容易，最好利用 Loopback 接口设置路由器的 ID；

EIGRP 的管理距离是 90；OSPF 的管理距离是 110；RIP 的管理距离是 120。因此当它们同时运行时，优先起作用的是 EIGRP。

直连路由的管理距离是 0，是最可靠、优先级最高的路由。默认路由的优先级最低。

课后作业

完成上面的模拟实训，将实训过程的截图按顺序粘贴到一个 Word 文件里并用适当的文字说明你对它的理解；总结本次实训所需要的主要命令及其作用，作为实训报告上交。

根据项目 01 中的路由器命名规则，务必把路由器命名为 RtIDa，其中 ID 为各自学号的最后两位数字。实训报告一律以"ID 姓名项目号.doc"为文件名命名，网络拓扑及其配置也以"ID 姓名项目号.pkt"为文件名保存并上交。例如，张三的 ID 为 03，他的文件名为"03 张三 08.doc"和"03 张三 08.pkt"。

思考题

当网络规模增大时，路由表的条目也将明显增多，这不仅会占用更多内存资源，在查找路由时还会消耗更多的 CPU 计算处理时间。有什么办法能减少路由表的条目吗？

在下一个项目中我们将学习有关的知识和技术。

项目 09　EIGRP 综合配置

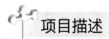

若你是某单位的网络管理员,你需要节约网络资源、提高网络效率,这包括合理规划子网和分配 IP 地址,利用路由汇总技术减少路由表的条目。

网络拓扑

网络拓扑如图 9-1 所示。

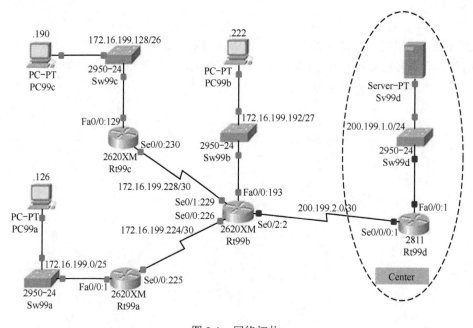

图 9-1　网络拓扑

学习目标

- 重点掌握 EIGRP 的配置方法;

项目 09 EIGRP 综合配置

- 重点掌握汇总路由的配置方法；
- 掌握查看路由的方法；
- 掌握调试网络的方法；
- 熟练掌握各接口的配置方法。

实训任务分解

① 前期准备。
② EIGRP 的配置与检验。
③ 汇总路由的配置与检验。
④ 模拟上网浏览。
⑤ 最佳静态路由配置。

知识点介绍

当网络规模增大时，路由表的条目也将明显增多。这不仅会占用更多内存资源，在查找路由时还会消耗更多的 CPU 计算处理时间。有什么办法能减少路由表的条目吗？

在合理规划子网和分配 IP 地址的基础上，利用路由汇总技术可极大地减少路由表的条目。

实训过程

实训环境在项目 08 的基础上进行了扩展，网络拓扑如图 9-1 所示。左边为原来的网络，模拟一个企业分部的网络；右边为新增的网络，模拟一个企业的网络中心，中心路由器 Rt99d 连接多个分部。（其他分部的网络与左边的类似，只是没画出来。）

也可把左边看成一个企业的网络，把右边看成因特网。ISP 的路由器 Rt99d 连接多个企业网。（其他企业的网络与左边的类似，只是没画出来。）

IP 地址和子网掩码等配置数据也已在网络拓扑中标明。

1. 前期准备

由于很多配置已在项目 08 中完成，故以下只列出新增部分的配置。

（1）构建如图 9-1 所示的网络拓扑

启动模拟器，打开 P08001.pkt 文件，然后再把它另存为文件名为 P09001.pkt 的文件。按图 9-1 扩展和连接网络。注意路由器 Rt99d 的型号为 2811，它可以增配四块 WIC-2T 模块。路由器 Rt99b 也要增配一块 WIC-1T 模块。

构建好网络拓扑后，单击【保存】按钮先将相关配置保存到 P09001.pkt 文件中备用，同时将其另存为文件名为 P09002.pkt 的文件，然后继续进行实训工作。

（2）完成 Rt99d 的基本配置

Router(config)#host Rt99d
Rt99d(config)#no ip domain-lookup
Rt99d(config)#enable secret 99secret
Rt99d(config)#service password-encry
Rt99d(config)#line con 0
Rt99d(config-line)#logg sync
Rt99d(config-line)#exec-timeout 0 0
Rt99d(config-line)#exit

注意：没有开启远程登录功能。

（3）完成 Rt99d 以太网接口的配置

Rt99d(config)#int fa0/0
Rt99d(config-if)#desc LAN link to Sw99d
Rt99d(config-if)#ip addr 200.199.1.1 255.255.255.0
Rt99d(config-if)#no shut

（4）完成 Rt99d 串行接口的配置

Rt99d(config-if)#int se0/0/0
Rt99d(config-if)#desc WAN link to Rt99b
Rt99d(config-if)#ip addr 200.199.2.1 255.255.255.252
Rt99d(config-if)#encap ppp
Rt99d(config-if)#ppp auth chap
Rt99d(config-if)#clock rate 2000000
Rt99d(config-if)#no shut
Rt99d(config-if)#exit
Rt99d(config)#username Rt99b password 99chap

注意：PPP 及其认证的配置。

（5）完成 Rt99b 串行接口的配置

Rt99b(config)#username Rt99d password 99chap
Rt99b(config)#int se0/2
Rt99b(config-if)#desc WAN link to Rt99d
Rt99b(config-if)#ip addr 200.199.2.2 255.255.255.252
Rt99b(config-if)#encap ppp
Rt99b(config-if)#ppp auth chap
Rt99b(config-if)#no shut

（6）完成服务器 IP 地址的配置

按下面的信息配置 Sv99d 的 IP 地址、子网掩码和默认网关。
- IP 地址：200.199.1.254；
- 子网掩码：255.255.255.0；
- 默认网关：200.199.1.1。

（7）测试直连网络的连通性

Rt99d#ping 200.199.2.2
!!!!!
Rt99d#ping 200.199.1.254
!!!!!

（8）查看 Rt99a 的路由信息

Rt99a# show ip route
Gateway of last resort is not set
172.16.0.0/16 is variably subnetted, 2 subnets, 2 masks
C 172.16.199.0/25 is directly connected, FastEthernet0/0
C 172.16.199.224/30 is directly connected, Serial0/0/0

（9）查看 Rt99b 的路由信息

Rt99b# show ip route
Gateway of last resort is not set
172.16.0.0/16 is variably subnetted, 3 subnets, 2 masks
C 172.16.199.192/27 is directly connected, FastEthernet0/0
C 172.16.199.224/30 is directly connected, Serial0/0
C 172.16.199.228/30 is directly connected, Serial0/1
200.199.2.0/24 is variably subnetted, 2 subnets, 2 masks
C 200.199.2.0/30 is directly connected, Serial0/2
C 200.199.2.1/32 is directly connected, Serial0/2

（10）查看 Rt99c 的路由信息

Rt99c# show ip route
Gateway of last resort is not set
172.16.0.0/16 is variably subnetted, 2 subnets, 2 masks
C 172.16.199.128/26 is directly connected, FastEthernet0/0
C 172.16.199.228/30 is directly connected, Serial0/0/0

（11）查看 Rt99d 的路由信息

Rt99d# show ip route

```
Gateway of last resort is not set
C   200.199.1.0/24 is directly connected, FastEthernet0/0
    200.199.2.0/24 is variably subnetted, 2 subnets, 2 masks
C   200.199.2.0/30 is directly connected, Serial0/0/0
C   200.199.2.2/32 is directly connected, Serial0/0/0
```

（12）保存配置

```
Rt99a# write
Rt99b# write
Rt99c# write
Rt99d# write
```

单击【保存】按钮先将相关配置保存到 P09002.pkt 文件中备用，再将其另存为文件名为 P09003.pkt 的文件后继续实训任务。

2. EIGRP 的配置与检验

（1）在 Rt99a 上配置 EIGRP

```
Rt99a(config)# router eigrp 333                        //启动 EIGRP 进程
Rt99a(config-router)# no auto-summary                  //取消自动汇总功能
Rt99a(config-router)# network 172.16.0.0               //通告网络
Rt99a(config-router)# passive-interface fa0/0          //被动接口
Rt99a(config-router)# end
```

（2）在 Rt99c 上配置 EIGRP

```
Rt99c(config)# router eigrp 333                        //启动 EIGRP 进程
Rt99c(config-router)# no auto-summary                  //取消自动汇总功能
Rt99c(config-router)# network 172.16.0.0               //通告网络
Rt99c(config-router)# passive-interface fa0/0          //被动接口
Rt99c(config-router)# end
```

（3）在 Rt99b 上配置 EIGRP

```
Rt99b(config)# router eigrp 333
Rt99b(config-router)# no auto-summary
Rt99b(config-router)# passive-int fa0/0
Rt99b(config-router)# network 200.199.2.0
Rt99b(config-router)# network 172.16.0.0
Rt99b(config-router)# end
```

（4）在 Rt99d 上配置 EIGRP

```
Rt99d(config)# router eigrp 333
Rt99d(config-router)# no auto-summary
Rt99d(config-router)# passive-int fa0/0
```

Rt99d(config-router)# network 200.199.1.0
Rt99d(config-router)# network 200.199.2.0
Rt99d(config-router)# end

（5）在 Rt99a 上查看路由信息

Rt99a# show ip route
Gateway of last resort is not set
172.16.0.0/16 is variably subnetted, 5 subnets, 4 masks
C 172.16.199.0/25 is directly connected, FastEthernet0/0
D 172.16.199.128/26 [90/2684416] via 172.16.199.226, 01:03:29, Serial0/0/0
D 172.16.199.192/27 [90/2172416] via 172.16.199.226, 01:03:29, Serial0/0/0
C 172.16.199.224/30 is directly connected, Serial0/0/0
D 172.16.199.228/30 [90/2681856] via 172.16.199.226, 01:03:29, Serial0/0/0
D 200.199.1.0/24 [90/2684416] via 172.16.199.226, 00:00:57, Serial0/0/0
200.199.2.0/30 is subnetted, 1 subnets
D 200.199.2.0 [90/2681856] via 172.16.199.226, 01:03:29, Serial0/0/0

（6）在 Rt99b 上查看路由信息

Rt99b# show ip route
Gateway of last resort is not set
172.16.0.0/16 is variably subnetted, 5 subnets, 4 masks
D 172.16.199.0/25 [90/2172416] via 172.16.199.225, 01:06:26, Serial0/0
D 172.16.199.128/26 [90/2172416] via 172.16.199.230, 01:06:26, Serial0/1
C 172.16.199.192/27 is directly connected, FastEthernet0/0
C 172.16.199.224/30 is directly connected, Serial0/0
C 172.16.199.228/30 is directly connected, Serial0/1
D 200.199.1.0/24 [90/2172416] via 200.199.2.1, 00:03:54, Serial0/2
200.199.2.0/24 is variably subnetted, 2 subnets, 2 masks
C 200.199.2.0/30 is directly connected, Serial0/2
C 200.199.2.1/32 is directly connected, Serial0/2

（7）在 Rt99c 上查看路由信息

Rt99c# show ip route
Gateway of last resort is not set
172.16.0.0/16 is variably subnetted, 5 subnets, 4 masks
D 172.16.199.0/25 [90/2684416] via 172.16.199.229, 01:07:40, Serial0/0/0
C 172.16.199.128/26 is directly connected, FastEthernet0/0
D 172.16.199.192/27 [90/2172416] via 172.16.199.229, 01:07:40, Serial0/0/0
D 172.16.199.224/30 [90/2681856] via 172.16.199.229, 01:07:40, Serial0/0/0
C 172.16.199.228/30 is directly connected, Serial0/0/0
D 200.199.1.0/24 [90/2684416] via 172.16.199.229, 00:05:08, Serial0/0/0
200.199.2.0/30 is subnetted, 1 subnets

D 200.199.2.0 [90/2681856] via 172.16.199.229, 01:07:40, Serial0/0/0

（8）在 Rt99d 上查看路由信息

Rt99d# show ip route
Gateway of last resort is not set
172.16.0.0/16 is variably subnetted, 5 subnets, 4 masks
D 172.16.199.0/25 [90/2684416] via 200.199.2.2, 00:06:22, Serial0/0/0
D 172.16.199.128/26 [90/2684416] via 200.199.2.2, 00:06:22, Serial0/0/0
D 172.16.199.192/27 [90/2172416] via 200.199.2.2, 00:06:22, Serial0/0/0
D 172.16.199.224/30 [90/2681856] via 200.199.2.2, 00:06:22, Serial0/0/0
D 172.16.199.228/30 [90/2681856] via 200.199.2.2, 00:06:22, Serial0/0/0
C 200.199.1.0/24 is directly connected, FastEthernet0/0
200.199.2.0/24 is variably subnetted, 2 subnets, 2 masks
C 200.199.2.0/30 is directly connected, Serial0/0/0
C 200.199.2.2/32 is directly connected, Serial0/0/0

注意：现在有 5 条动态路由。

（9）在 Rt99a 上查看路由协议相关信息

Rt99a# show ip proto
Routing Protocol is "eigrp 333 "
……

（10）在 Rt99b 上查看路由协议相关信息

Rt99b# show ip proto
Routing Protocol is "eigrp 333 "
……

（11）在 Rt99c 上查看路由协议相关信息

Rt99c# show ip proto
Routing Protocol is "eigrp 333 "
……

（12）在 Rt99d 上查看路由协议相关信息

Rt99d# show ip proto
Routing Protocol is "eigrp 333 "
……

（13）在 Rt99a 上测试网络的连通性

Rt99a#ping 172.16.199.190
!!!!!
Rt99a#ping 172.16.199.222

项目 09　EIGRP 综合配置

!!!!!
Rt99a#ping 200.199.1.254
!!!!!

（14）在 Rt99d 上测试网络连通性

Rt99d#ping 172.16.199.126
.!!!!
Rt99d#ping 172.16.199.190
!!!!!
Rt99d#ping 172.16.199.222
!!!!!

（15）保存配置

Rt99a# write
Rt99b# write
Rt99c# write
Rt99d# write

单击【保存】按钮先将相关配置保存到 P09003.pkt 文件中备用，再将其另存为文件名为 P09004.pkt 的文件，然后继续进行实训工作。

3. 汇总路由的配置与检验

在图 9-1 中，只有 Rt99b 向 Rt99d 通告的路由适合汇总。因为左边的 5 个子网是由一个网络 172.16.199.0/24 划分出来的。因此，汇总路由只在路由器 Rt99b 的 Se0/2 接口上配置。

完成配置后请仔细观察路由信息并与之前的路由信息进行对比，然后重新全面检验整个网络的连通性，最后要记得保存配置。

（1）在 Rt99b 上配置汇总路由

Rt99b(config)# int se0/2
Rt99b(config-if)# ip summary-address eigrp 333 172.16.199.0 255.255.255.0
Rt99b(config-if)# end

（2）在 Rt99d 上查看路由信息

Rt99d# show ip route
Gateway of last resort is not set
172.16.0.0/24 is subnetted, 1 subnets
D 172.16.199.0 [90/2172416] via 200.199.2.2, 00:17:37, Serial0/0/0
C 200.199.1.0/24 is directly connected, FastEthernet0/0
200.199.2.0/24 is variably subnetted, 2 subnets, 2 masks
C 200.199.2.0/30 is directly connected, Serial0/0/0
C 200.199.2.2/32 is directly connected, Serial0/0/0

注意：现在只有 1 条动态路由了（之前有 5 条）。

在 Rt99b 上配置汇总路由前，Rt99d 有 5 条动态路由；在 Rt99b 上配置汇总路由后，Rt99d 就只有 1 条动态路由。如果网络中心的路由器 Rt99d 连接 10 个类似的分部网络，则在进行路由汇总前将有 50 条动态路由，在进行路由汇总后便只有 10 条了。一个小小的改变，使路由表项大量减少。这样累积的结果对改善大型网络的性能将产生巨大的影响。

但是路由汇总依赖于恰当的 IP 地址分配，这就是总部要给分部指定 IP 地址的原因。作为总部高层的网络管理员，你必须对整个网络的构建进行全面的考虑和规划，使网络结构合理、性能高效、维护方便。

（3）在 Rt99c 上测试网络连通性

```
Rt99c#ping 172.16.199.126
!!!!!
Rt99c#ping 172.16.199.222
!!!!!
Rt99c#ping 200.199.1.254
!!!!!
```

（4）在 Rt99d 上测试网络连通性

```
Rt99d#ping 172.16.199.222
!!!!!
Rt99d#ping 172.16.199.190
!!!!!
Rt99d#ping 172.16.199.126
!!!!!
```

测试结果说明，Rt99d 上虽然只有一条动态路由，但整个网络是全连通的。

（5）保存配置

```
Rt99b# write
Rt99d# write
```

单击【保存】按钮先将相关配置保存到 P09004.pkt 文件中备用，再将其另存为文件名为 P09005.pkt 的文件，然后继续进行实训工作。

4．模拟上网浏览

在图 9-1 中的服务器 Sv99d 上配置和启动 HTTP 和 DNS 服务，再在各台 PC 上配置 DNS 服务器的 IP 地址，便可模拟上网浏览。

（1）在 Sv99d 上配置和启动 HTTP（如图 9-2 所示）

（2）在 Sv99d 上配置和启动 DNS（如图 9-3 所示）

项目 09　EIGRP 综合配置

图 9-2　在 Sv99d 上配置和启动 HTTP

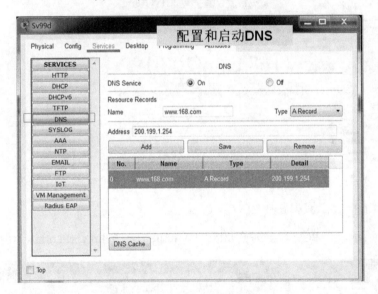

图 9-3　在 Sv99d 上配置和启动 DNS

（3）配置 PC 的 IP 地址

在各台 PC 上按如下信息配置 IP 地址、子网掩码和默认网关。

① PC99a。

- IP 地址：172.16.199.126；
- 子网掩码：255.255.255.128；
- 默认网关：172.16.199.1；
- DNS IP 地址：200.199.1.254。

② PC99b。

- IP 地址：172.16.199.222；
- 子网掩码：255.255.255.224；
- 默认网关：172.16.199.193；
- DNS IP 地址：200.199.1.254。

③ PC99c。
- IP 地址：172.16.199.190；
- 子网掩码：255.255.255.192；
- 默认网关：172.16.199.129；
- DNS IP 地址：200.199.1.254。

（4）在 PC99a 上上网浏览

如图 9-4 所示，在 PC99a 上浏览 http://www.168.com，访问成功！

图 9-4　在 PC99a 上浏览 http://www.168.com

（5）在 PC99b 上上网浏览

如图 9-5 所示，在 PC99b 上浏览 http://www.168.com/helloworld.html，访问成功！

图 9-5　在 PC99b 上浏览 http://www.168.com/helloworld.html

（6）在 PC99c 上上网浏览

如图 9-6 所示，在 PC99c 上浏览 http://www.168.com/image.html，访问成功！

项目 09　EIGRP 综合配置

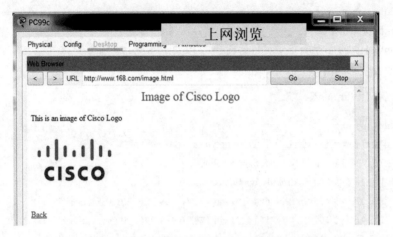

图 9-6　在 PC99c 上浏览 http://www.168.com/image.html

上面的测试结果说明，分部的每台主机都已能访问总部的服务器 Sv99d 上的网站了。而且能用域名访问，说明 DNS 域名解析服务也正常。

（7）保存文件

单击【保存】按钮先将相关配置保存到 P09005.pkt 文件中备用，再将其另存为文件名为 P09006.pkt 的文件，然后继续进行实训工作。

在图 9-1 中，如果右边是因特网，Rt99d 是 ISP 的路由器，企业网和因特网不会运行同一个路由协议（即使运行同一个路由协议，企业网和因特网之间也不许动态交换路由信息），那么在边界路由器 Rt99b 和 Rt99d 上如何配置路由呢？

为了禁止路由信息跨越边界传输，可将边界路由器的对外接口配置为被动接口。为了保持企业网和因特网之间仍能通信，可在边界路由器上配置静态路由或默认路由。

（8）修改 Rt99d 的配置

```
Rt99d(config)# ip route 172.16.199.0 255.255.255.0 200.199.2.2
Rt99d(config)# router eigrp 333
Rt99d(config-router)# redistribute static metric 1 20000 255 200 1500
//路由协议的迁移或重分配：带宽、延迟、可靠性、MTU
Rt99d(config-router)# passive-int se0/0/0
Rt99d(config-router)# end
```

（9）修改 Rt99b 的配置

```
Rt99b(config)# ip route 0.0.0.0 0.0.0.0 200.199.2.1
Rt99b(config)# router eigrp 333
Rt99b(config-router)# redistribute static metric 1 20000 255 200 1500
//路由协议的迁移或重分配：带宽、延迟、可靠性、MTU
Rt99b(config-router)# passive-int se0/2
```

Rt99b(config-router)#exit
Rt99b(config)# int se0/2
Rt99b(config-if)# no ip summary-address eigrp 333 172.16.199.0 255.255.255.0
Rt99b(config-if)# end

(10) 查看 Rt99a 的路由表

Rt99a# show ip route
Gateway of last resort is 172.16.199.226 to network 0.0.0.0
    172.16.0.0/16 is variably subnetted, 5 subnets, 4 masks
C   172.16.199.0/25 is directly connected, FastEthernet0/0
D   172.16.199.128/26 [90/2684416] via 172.16.199.226, 00:00:57, Serial0/0/0
D   172.16.199.192/27 [90/2172416] via 172.16.199.226, 00:00:57, Serial0/0/0
C   172.16.199.224/30 is directly connected, Serial0/0/0
D   172.16.199.228/30 [90/2681856] via 172.16.199.226, 00:00:57, Serial0/0/0
    200.199.2.0/30 is subnetted, 1 subnets
D   200.199.2.0 [90/2681856] via 172.16.199.226, 00:00:57, Serial0/0/0
D*EX 0.0.0.0/0 [170/2565632000] via 172.16.199.226, 00:00:57, Serial0/0/0
//请注意最后一条

(11) 查看 Rt99b 的路由表

Rt99b# show ip route
Gateway of last resort is 200.199.2.1 to network 0.0.0.0
    172.16.0.0/16 is variably subnetted, 5 subnets, 4 masks
D   172.16.199.0/25 [90/2172416] via 172.16.199.225, 00:01:46, Serial0/0
D   172.16.199.128/26 [90/2172416] via 172.16.199.230, 00:01:46, Serial0/1
C   172.16.199.192/27 is directly connected, FastEthernet0/0
C   172.16.199.224/30 is directly connected, Serial0/0
C   172.16.199.228/30 is directly connected, Serial0/1
    200.199.2.0/24 is variably subnetted, 2 subnets, 2 masks
C   200.199.2.0/30 is directly connected, Serial0/2
C   200.199.2.1/32 is directly connected, Serial0/2
S*   0.0.0.0/0 [1/0] via 200.199.2.1

(12) 查看 Rt99c 的路由表

Rt99c# show ip route
Gateway of last resort is 172.16.199.229 to network 0.0.0.0
    172.16.0.0/16 is variably subnetted, 5 subnets, 4 masks
D   172.16.199.0/25 [90/2684416] via 172.16.199.229, 00:02:53, Serial0/0/0
C   172.16.199.128/26 is directly connected, FastEthernet0/0
D   172.16.199.192/27 [90/2172416] via 172.16.199.229, 00:02:53, Serial0/0/0
D   172.16.199.224/30 [90/2681856] via 172.16.199.229, 00:02:53, Serial0/0/0

C   172.16.199.228/30 is directly connected, Serial0/0/0
         200.199.2.0/30 is subnetted, 1 subnets
D   200.199.2.0 [90/2681856] via 172.16.199.229, 00:02:53, Serial0/0/0
D*EX 0.0.0.0/0 [170/2565632000] via 172.16.199.229, 00:02:53, Serial0/0/0
//请注意最后一条

（13）查看 Rt99d 的路由表

Rt99d# show ip route
Gateway of last resort is not set
         172.16.0.0/24 is subnetted, 1 subnets
S   172.16.199.0 [1/0] via 200.199.2.2
C   200.199.1.0/24 is directly connected, FastEthernet0/0
         200.199.2.0/24 is variably subnetted, 2 subnets, 2 masks
C   200.199.2.0/30 is directly connected, Serial0/0/0
C   200.199.2.2/32 is directly connected, Serial0/0/0

（14）在 PC99c 上上网浏览

如图 9-7 所示，在 PC99c 上浏览 http://www.168.com/index.html，访问成功！

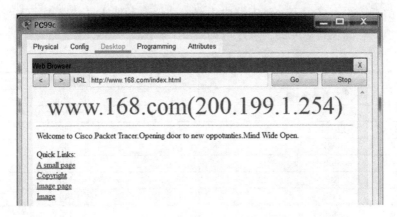

图 9-7　在 PC99c 上浏览 http://www.168.com/index.html

（15）保存配置

Rt99a# write
Rt99b# write
Rt99c# write
Rt99d# write

单击【保存】按钮先将相关配置保存到 P09006.pkt 文件中备用，再将其另存为文件名为 P09007.pkt 的文件。

5．最佳静态路由配置

路由表项太多会消耗大量 CPU 的计算资源，通过配置适当的汇总路由和默认路由可以大大减少路由表项。

静态路由也可进行路由汇总。譬如 Rt99d 既然模拟因特网,显然不适合配置默认路由,但可以配置汇总路由。对于规模不大、结构稳定的网络,静态路由是否更好?

对于图 9-1 所示网络,若改成配置静态路由,则如何实现最佳配置?这里的所谓最佳就是每台路由器的路由表项最少。

这可在文件 P09002.pkt 保存内容的基础上进行。请打开文件 P09002.pkt,把它另存为文件名为 P09008.pkt 的文件,然后再完成自认最佳的静态路由配置。

学习总结

路由协议与 IP 和 ICMP 都处于第三层,即网络层。路由协议为 IP 提供服务,即为 IP 转发数据包提供正确的路由。

为了获得正确的路由信息,各路由器之间需要互相交流学习,这需要消耗一定的带宽,如何提高路由信息交换的效率或降低路由信息交换占用的带宽是选用和配置路由协议时需要考虑的重要问题。

动态路由的收敛(从开始互相学习到保持稳定状态)都需要一定的时间。

静态路由直接写入路由表,动态路由由路由协议自动选出最佳路由写入路由表。

常见路由信息源及其对应的管理距离如表 9-1 所示。

表 9-1　常见路由信息源及其对应的管理距离

路由信息源	默认管理距离
Connected(直连路由)	0
Static(静态路由)	1
EIGRP	90
IGRP	100
OSPF	110
RIP	120
EGP	140
未知	255

管理距离越小优先级越高,默认路由的优先级最低。

课后作业

完成上面的模拟实训,然后改为静态路由的最佳配置。

将实训过程的截图按顺序粘贴到一个 Word 文件里并用适当的文字说明你对它的理解;总结本次实训所需要的主要命令及其作用,作为实训报告上交。

实训报告一律以"ID 姓名项目号.doc"为文件名命名,网络拓扑及其配置也以"ID 姓名项目号.pkt"为文件名保存并上交。例如,张三的 ID 为 03,他的文件名为"03 张三 09.doc"和"03 张三 09.pkt"。

项目 09　EIGRP 综合配置

思考题

对于结构稳定的小型网络，静态路由是否更好？

在远程连接中，我们至今都采用串行连接，其封装协议为 HDLC 或 PPP。如果租用帧中继网络，则又如何进行远程连接和配置？

在下一个项目中我们将学习有关的知识和技术。

项目 10　帧中继与子接口配置

项目描述

某公司在北京、上海、重庆和广州有四个分部，现决定租用中国电信的帧中继网络把各分部的局域网连接起来组成一个广域网，网络拓扑如图 10-1 所示。

假设你是该公司的网络管理员，如何构建该网络？

如果你想先用模拟器来模拟该网络的构建，怎么做？

网络拓扑

网络拓扑如图 10-1 所示。

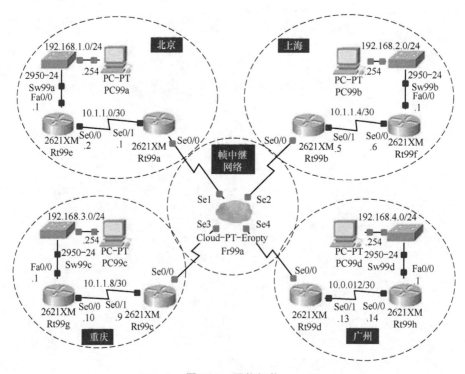

图 10-1　网络拓扑

项目 10　帧中继与子接口配置

学习目标

- 重点掌握帧中继和子接口的配置方法；
- 掌握模拟帧中继网络的方法；
- 掌握调试网络的方法；
- 熟练掌握各接口的配置方法。

实训任务分解

① 模拟帧中继网络。
② 完成帧中继与子接口的相关配置。
③ 完成各分部网络的配置。

知识点介绍

广域网（WAN）通常是通过租用电信运营商提供的数据链路，将位于各地的多个局域网（LAN）连接起来而构成的一个通信网络。

广域网技术有帧中继、DDN、X.25、ISDN、xDSL、各种专线等，其性能和价格千差万别，其中专线、帧中继、xDSL 是目前最常用的技术。

我们前面学的串行连接就是专线连接，其封装协议为 HDLC 或 PPP。

在本项目中我们将学习帧中继的配置及其相关知识。

实训过程

1. 模拟帧中继网络

实际工作中可租用电信运营商的帧中继网络，但在实训环境中没有真实的帧中继网络，怎么办？在实训环境中我们可以先模拟一个帧中继网络，具体模拟步骤和方法如下所述。

① 如图 10-2 所示，首先选择 WAN Emulation 云图，然后选择 Cloud-PT-Empty 云图。
② 如图 10-3 所示，把 Cloud-PT-Empty 云图拖放到中间。
③ 如图 10-4 所示，改名后双击 Cloud-PT-Empty 云图。
④ 如图 10-5 所示，关闭电源开关。
⑤ 如图 10-6 所示，先添加 4 个 PT-CLOUD-NM-1S 模块，然后加电。

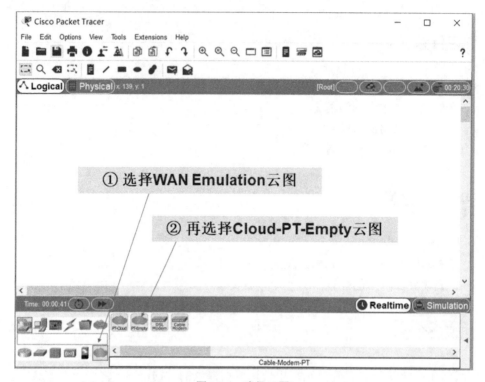

图 10-2　选择云图

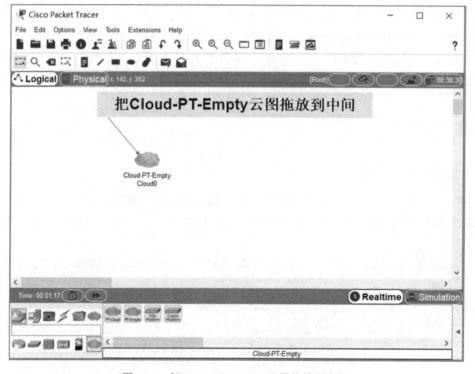

图 10-3　把 Cloud-PT-Empty 云图拖放到中间

项目 10　帧中继与子接口配置

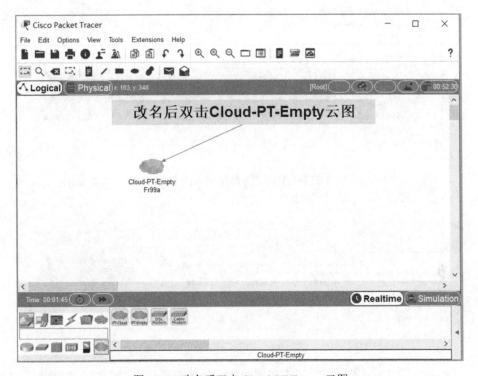

图 10-4　改名后双击 Cloud-PT-Empty 云图

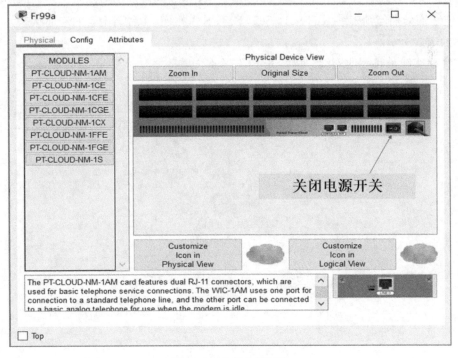

图 10-5　关闭电源开关

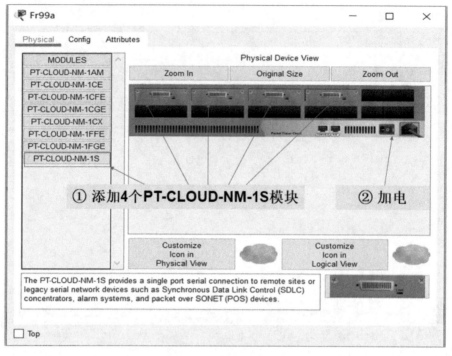

图 10-6　添加模块后加电

主干拓扑如图 10-7 所示。为了在云图内模拟帧中继网络，首先将 Se1 分别与 Se2 和 Se3 连通，Se4 分别与 Se2 和 Se3 连通，如图 10-7 中云内部的虚线所示。然后增加 4 台路由器，使它们分别与帧中继云的 Se1、Se2、Se3、Se4 相连，4 台路由器的连通状态如图 10-7 中外部的虚线所示。

注释：实线是实际的连线，虚线是 PVC（永久虚链路）。

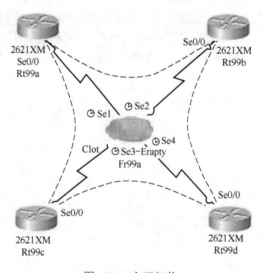

图 10-7　主干拓扑

⑥ 如图 10-8 所示，配置 Se1 的帧中继 DLCI。
DLCI：Data Link Connection Identifier，数据链路连接标识。

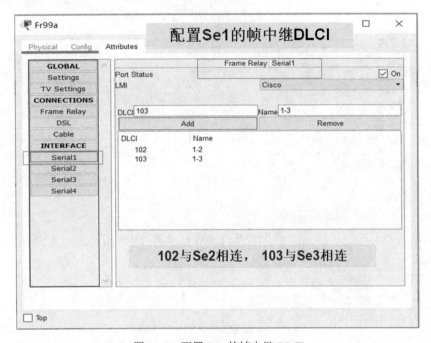

图 10-8　配置 Se1 的帧中继 DLCI

⑦ 如图 10-9 所示，配置 Se2 的帧中继 DLCI。

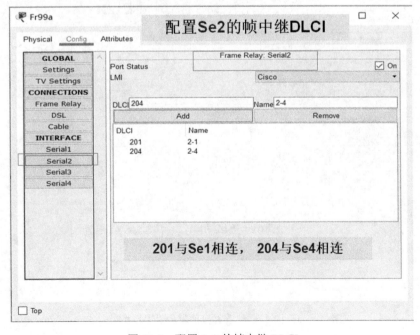

图 10-9　配置 Se2 的帧中继 DLCI

⑧ 如图 10-10 所示,配置 Se3 的帧中继 DLCI。

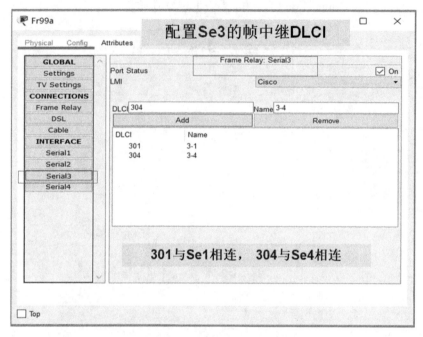

图 10-10　配置 Se3 的帧中继 DLCI

⑨ 如图 10-11 所示,配置 Se4 的帧中继 DLCI。

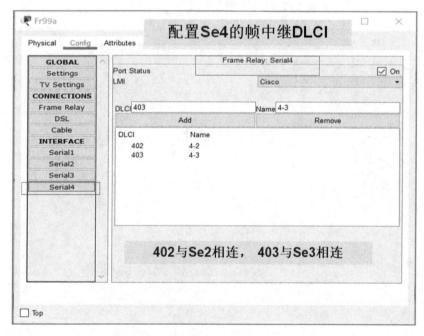

图 10-11　配置 Se4 的帧中继 DLCI

⑩ 在 Fr99a 上配置帧中继连接:添加 Se1 到 Se2 和 Se3 的连接链路,如图 10-12 所示。

项目 10　帧中继与子接口配置

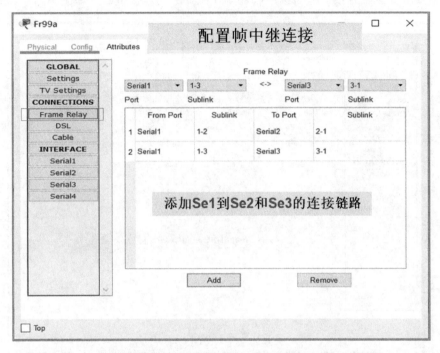

图 10-12　配置帧中继连接：添加 Se1 到 Se2 和 Se3 的连接链路

⑪ 在 Fr99a 上配置帧中继连接：添加 Se2 到 Se4 的连接链路，如图 10-13 所示。

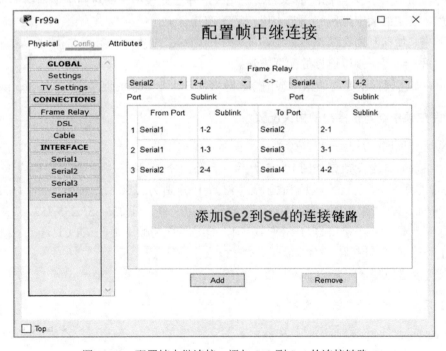

图 10-13　配置帧中继连接：添加 Se2 到 Se4 的连接链路

⑫ 在 Fr99a 上配置帧中继连接：添加 Se3 到 Se4 的连接链路，如图 10-14 所示。

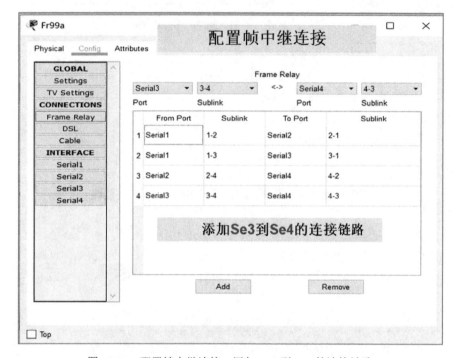

图 10-14 配置帧中继连接：添加 Se3 到 Se4 的连接链路

至此，图 10-7 中云内部的虚线连接已完成，即已完成用云图对帧中继网络的模拟。

注意：真实的帧中继网络通常由电信运营商提供。

接下来是如何利用在模拟器上搭建的帧中继网络来连接 4 台远程路由器。这有多种方案，我们只介绍用子接口连接的配置方法。

思考：如何模拟全连通的帧中继网络，即增加 Se1 与 Se4 的连接、Se2 与 Se3 的连接？

2. 完成帧中继与子接口的相关配置

在实际建网时，从这一步开始才是真正需要做的工作。

帧中继的有关配置数据如图 10-15 所示。4 台路由器之间通过帧中继网络构成 4 个子网，如外边的虚线所示，子网地址和掩码（位数）分别为 200.1.1.16/30、200.1.1.20/30、200.1.1.24/30、200.1.1.28/30。其中，子网掩码都是 30 位，即 255.255.255.252。

为了易于辨识，子接口号与子网号相同，冒号后面为帧中继的 DLCI 号。子接口号采用 DLCI 号可能更易识别。

4 台路由器的基本配置如前，故以下省略了。

（1）完成 Rt99a 的相关配置

```
Rt99a(config)#interface Se0/0
Rt99a(config-if)#no ip address
Rt99a(config-if)#encapsulation frame-relay
```

项目 10 帧中继与子接口配置

```
Rt99a(config-if)#interface Se0/0.16 point-to-point
Rt99a(config-subif)#desc WAN link to Rt99b
Rt99a(config-subif)#ip address 200.1.1.17 255.255.255.252
Rt99a(config-subif)#frame-relay interface-dlci 102
Rt99a(config-subif)#interface Serial0/0.20 point-to-point
Rt99a(config-subif)#desc WAN link to Rt99c
Rt99a(config-subif)#ip address 200.1.1.21 255.255.255.252
Rt99a(config-subif)#frame-relay interface-dlci 103
Rt99a(config-subif)#interface Serial0/0
Rt99a(config-if)#no shutdown
```

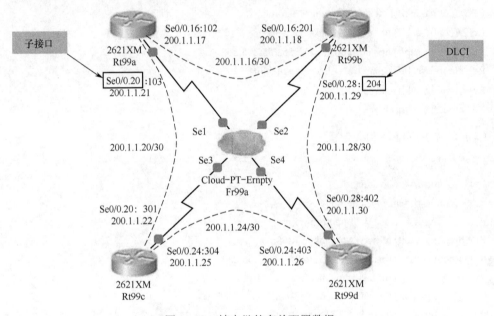

图 10-15 帧中继的有关配置数据

（2）完成 Rt99b 的相关配置

```
Rt99b(config)#int se0/0
Rt99b(config-if)#no ip addr
Rt99b(config-if)#encap frame-relay
Rt99b(config-if)#int se0/0.16 point-to-point
Rt99b(config-subif)#desc WAN Link to Rt99a
Rt99b(config-subif)#ip addr 200.1.1.18 255.255.255.252
Rt99b(config-subif)#frame-relay interface-dlci 201
Rt99b(config-subif)#int se0/0.28 point-to-point
Rt99b(config-subif)#desc WAN Link to Rt99d
Rt99b(config-subif)#ip addr 200.1.1.29 255.255.255.252
Rt99b(config-subif)#frame-relay interface-dlci 204
Rt99b(config-subif)#int se0/0
```

117

Rt99b(config-if)#no shut

（3）完成 Rt99c 的相关配置

Rt99c(config)#int se0/0
Rt99c(config-if)#no ip addr
Rt99c(config-if)#encap frame-relay
Rt99c(config-if)#int se0/0.20 point-to-point
Rt99c(config-subif)#desc WAN Link to Rt99a
Rt99c(config-subif)#ip addr 200.1.1.22 255.255.255.252
Rt99c(config-subif)#frame-relay interface-dlci 301
Rt99c(config-subif)#int se0/0.24 point-to-point
Rt99c(config-subif)#desc WAN Link to Rt99d
Rt99c(config-subif)#ip addr 200.1.1.25 255.255.255.252
Rt99c(config-subif)#frame-relay interface-dlci 304
Rt99c(config-subif)#int se0/0
Rt99c(config-if)#no shut

（4）完成 Rt99d 的相关配置

Rt99d(config)#int se0/0
Rt99d(config-if)#no ip addr
Rt99d(config-if)#encap frame-relay
Rt99d(config-if)#int se0/0.24 point-to-point
Rt99d(config-subif)#desc WAN Link to Rt99c
Rt99d(config-subif)#ip addr 200.1.1.26 255.255.255.252
Rt99d(config-subif)#frame-relay interface-dlci 403
Rt99d(config-subif)#int se0/0.28 point-to-point
Rt99d(config-subif)#desc WAN Link to Rt99b
Rt99d(config-subif)#ip addr 200.1.1.30 255.255.255.252
Rt99d(config-subif)#frame-relay interface-dlci 402
Rt99d(config-subif)#int se0/0
Rt99d(config-if)#no shut

（5）在 Rt99a 上测试直连网络的连通性

Rt99a#ping 200.1.1.18
!!!!!
Rt99a#ping 200.1.1.22
!!!!!
Rt99a#ping 200.1.1.25
……

（6）在 Rt99b 上测试直连网络的连通性

Rt99b#ping 200.1.1.17

!!!!!
Rt99b#ping 200.1.1.30
!!!!!
Rt99b#ping 200.1.1.21
......

（7）在 Rt99c 上测试直连网络的连通性

Rt99c#ping 200.1.1.21
!!!!!
Rt99c#ping 200.1.1.26
!!!!!
Rt99c#ping 200.1.1.30
......

（8）在 Rt99d 上测试直连网络的连通性

Rt99d#ping 200.1.1.25
!!!!!
Rt99d#ping 200.1.1.29
!!!!!
Rt99d#ping 200.1.1.22
......

（9）保存配置

对网络进行测试，如果相邻路由器都能连通，则说明帧中继和子接口的配置正确；否则，要找出错误并纠正。在验证配置正确后，要用以下命令保存配置：

copy run start

千万要记得保存相关网络拓扑！请先将其保存为文件名为 P10001.pkt 的文件备用，然后再将其另存为文件名为 P10002.pkt 的文件。

3. 完成各分部网络的配置

在图 10-15 的基础上进行扩展，将各分部网络连接好，网络拓扑如图 10-16 所示。它与图 10-1 中的网络拓扑相同。IP 地址和子网掩码等配置数据已在图 10-16 中标明。下面只列出北京分部网络的配置，其他分部网络的配置类似。上海、重庆、广州分部网络的串行线路请选配不同的时钟频率。

（1）完成 Rt99e 的基本配置

Router(config)#host Rt99e
Rt99e(config)#enable secret 99secret
Rt99e(config)#line con 0
Rt99e(config-line)#logg sync

Rt99e(config-line)#exec-timeout 0 0
Rt99e(config-line)#line vty 0 15
Rt99e(config-line)#password 99vty015
Rt99e(config-line)#login
Rt99e(config-line)#logg sync
Rt99e(config-line)#exec-timeout 0 0
Rt99e(config-line)#exit
Rt99e(config)#service password-encryption
Rt99e(config)#no ip domain-lookup

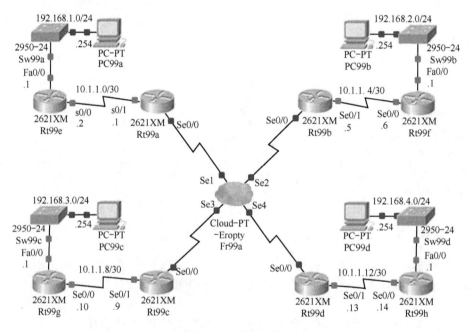

图 10-16　网络拓扑

（2）完成 Rt99a 的接口配置

Rt99a(config)#int se0/1
Rt99a(config-if)#desc WAN Link to Rt99e
Rt99a(config-if)#ip addr 10.1.1.1 255.255.255.252
Rt99a(config-if)#no shut
Rt99a(config-if)#end

（3）完成 Rt99e 的接口配置

Rt99e(config)#int se0/0
Rt99e(config-if)#desc WAN Link to Rt99a
Rt99e(config-if)#ip addr 10.1.1.2 255.255.255.252
Rt99e(config-if)#clock rate 2000000

项目 10　帧中继与子接口配置

Rt99e(config-if)#no shut
Rt99e(config-if)#int fa0/0
Rt99e(config-if)#desc LAN Link to Sw99a
Rt99e(config-if)#ip addr 192.168.1.1 255.255.255.0
Rt99e(config-if)#no shut

（4）配置 PC 的 IP 地址

在 PC99a 上配置 IP 地址、子网掩码和默认网关，然后单击【保存】按钮，保存已有配置。

- PC99a 的 IP 地址：192.168.1.254；
- PC99a 的子网掩码：255.255.255.0；
- PC99a 的默认网关：192.168.1.1。

（5）在 Rt99e 上测试网络的连通性

Rt99e#ping 192.168.1.254
.!!!!
Rt99e#ping 10.1.1.1
!!!!!

（6）保存配置

在实际工作中，完成网络的配置和调试并确认路由器的配置正确后，都必须保存配置，否则，路由器重启后又要重新配置和调试。为了养成良好的习惯，每次完成模拟配置和调试后，保存每台路由器的配置，保存配置的命令为 copy run start。

将各分部的网络都配置好后保存好相关网络拓扑，先单击【保存】按钮保存相关网络拓扑，然后再将其另存为文件名为 P10003.pkt 的文件。

学习总结

路由器的串行接口在与帧中继网络连接时，可以逻辑地划分为多个虚拟的子接口，每个子接口可单独与一个远端子接口实现点对点相连，从而使子接口如专线一样使用。

子接口的编号可以是 $1\sim(2^{32}-1)$ 范围内的整数。

课后作业

完成上面的模拟实训，将实训过程的截图按顺序粘贴到一个 Word 文件里并用适当的文字说明你对它的理解；总结本次实训所需要的主要命令及其作用，作为实训报告上交。

实训报告一律以"ID 姓名项目号.doc"为文件名命名，网络拓扑及其配置也以"ID 姓名项目号.pkt"为文件名保存并上交。例如，张三的 ID 为 03，他的文件名为"03 张三 10.doc"和"03 张三 10.pkt"。

思考题

　　大型网络中不仅有不同厂商的网络设备，而且还都具有一定的层次结构，多区域的 OSPF 恰好与之相适应，因此，OSPF 得到了广泛的应用。

　　前面我们学习了单区域 OSPF 的配置，那么如何配置多区域 OSPF 呢？

　　在下一个项目中我们将学习有关的知识和技术。

项目 11　OSPF 综合配置

项目描述

某公司在北京、上海、重庆和广州有四个分部，现决定租用中国电信的帧中继网络将各分部的局域网连接起来组成一个广域网。

假设你是该公司的网络管理员，你决定为该网络配置多区域 OSPF，该如何实现？

网络拓扑

网络拓扑如图 11-1 所示。

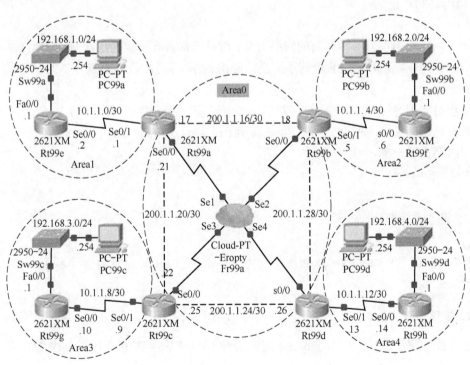

图 11-1　网络拓扑

学习目标

- 重点掌握多区域 OSPF 的配置方法；
- 掌握 OSPF 各种信息的查看方法；
- 掌握查看路由的方法；
- 掌握调试网络的方法。

实训任务分解

① 配置各路由器的 ID。
② 配置多区域 OSPF。
③ 查看 OSPF 各种信息。
④ 完成网络调试验收。
⑤ 模拟上网浏览。

知识点介绍

大型网络中不仅有不同厂商的网络设备，而且还都具有一定的层次结构，多区域 OSPF 恰好与此相适应，因此，OSPF 得到了广泛的应用。在本项目中我们将重点学习多区域 OSPF 的配置。

本项目实训在项目 10 的基础上进行，实训环境和配置数据如图 11-1 所示，其中网络拓扑连接、路由器接口和 PC 等基础性的配置已在项目 10 中完成，执行以下命令可以查看已完成的路由器配置：

Rt99a# show run

实训过程

把项目 10 的 P10003.pkt 文件另存为文件名为 P11001.pkt 的文件备用，然后打开 P11001.pkt 文件。

1. 配置各路由器的 ID

通过配置每个路由器的 loopback0 接口的 IP 地址来设置路由器 ID，如表 11-1 所示。

表 11-1 设置路由器 ID

Router	Router ID	Router	Router ID
Rt99a	10.1.1.21	Rt99e	10.1.1.25

续表

Router	Router ID	Router	Router ID
Rt99b	10.1.1.22	Rt99f	10.1.1.26
Rt99c	10.1.1.23	Rt99g	10.1.1.27
Rt99d	10.1.1.24	Rt99h	10.1.1.28

注意：loopback0 接口的子网掩码为 255.255.255.255。

（1）配置路由器 Rt99a 的 ID

Rt99a(config)#int loopback0
Rt99a(config-if)#ip addr 10.1.1.21 255.255.255.255
Rt99a(config-if)#end

（2）配置路由器 Rt99b 的 ID

Rt99b(config)#int loopback0
Rt99b(config-if)#ip addr 10.1.1.22 255.255.255.255
Rt99b(config-if)#end

（3）配置路由器 Rt99c 的 ID

Rt99c(config)#int loopback0
Rt99c(config-if)#ip addr 10.1.1.23 255.255.255.255
Rt99c(config-if)#end

（4）配置路由器 Rt99d 的 ID

Rt99d(config)#int loopback0
Rt99d(config-if)#ip addr 10.1.1.24 255.255.255.255
Rt99d(config-if)#end

（5）配置路由器 Rt99e 的 ID

Rt99e(config)#int loopback0
Rt99e(config-if)#ip addr 10.1.1.25 255.255.255.255
Rt99e(config-if)#end

（6）配置路由器 Rt99f 的 ID

Rt99f(config)#int loopback0
Rt99f(config-if)#ip addr 10.1.1.26 255.255.255.255
Rt99f(config-if)#end

（7）配置路由器 Rt99g 的 ID

Rt99g(config)#int loopback0
Rt99g(config-if)#ip addr 10.1.1.27 255.255.255.255
Rt99g(config-if)#end

（8）配置路由器 Rt99h 的 ID

Rt99h(config)#int loopback0

Rt99h(config-if)#ip addr 10.1.1.28 255.255.255.255
Rt99h(config-if)#end

2. 配置多区域 OSPF

在多区域 OSPF 中，Area0 为骨干区域，其他区域都必须与它相连。连接 Area0 和其他区域的路由器被称为区域边界路由器（ABR）。Rt99a、Rt99b、Rt99c、Rt99d 都是 ABR。

（1）在路由器 Rt99a 上配置 OSPF

```
Rt99a(config)#router ospf 100
Rt99a(config-router)#network 200.1.1.16 0.0.0.3 area 0
Rt99a(config-router)#network 200.1.1.20 0.0.0.3 area 0
Rt99a(config-router)#network 10.1.1.0 0.0.0.3 area 1
Rt99a(config-router)#end
```

（2）在路由器 Rt99b 上配置 OSPF

```
Rt99b(config)#router ospf 200
Rt99b(config-router)#network 200.1.1.16 0.0.0.3 area 0
Rt99b(config-router)#network 200.1.1.28 0.0.0.3 area 0
Rt99b(config-router)#network 10.1.1.4 0.0.0.3 area 2
Rt99b(config-router)#end
```

（3）在路由器 Rt99c 上配置 OSPF

```
Rt99c(config)#router ospf 300
Rt99c(config-router)#network 200.1.1.20 0.0.0.3 area 0
Rt99c(config-router)#network 200.1.1.24 0.0.0.3 area 0
Rt99c(config-router)#network 10.1.1.8 0.0.0.3 area 3
Rt99c(config-router)#end
```

（4）在路由器 Rt99d 上配置 OSPF

```
Rt99d(config)#router ospf 400
Rt99d(config-router)#network 200.1.1.24 0.0.0.3 area 0
Rt99d(config-router)#network 200.1.1.28 0.0.0.3 area 0
Rt99d(config-router)#network 10.1.1.12 0.0.0.3 area 4
Rt99d(config-router)#end
```

（5）在路由器 Rt99e 上配置 OSPF

```
Rt99e(config)#router ospf 101
Rt99e(config-router)#network 10.1.1.0 0.0.0.3 area 1
Rt99e(config-router)#network 192.168.1.0 0.0.0.255 area 1
Rt99e(config-router)#passive-interface fa0/0
Rt99e(config-router)#end
```

项目 11　OSPF 综合配置

（6）在路由器 Rt99f 上配置 OSPF

Rt99f(config)#router ospf 202
Rt99f(config-router)#network 10.1.1.4 0.0.0.3 area 2
Rt99f(config-router)#network 192.168.2.0 0.0.0.255 area 2
Rt99f(config-router)#passive-interface fa0/0
Rt99f(config-router)#end

（7）在路由器 Rt99g 上配置 OSPF

Rt99g(config)#router ospf 303
Rt99g(config-router)#network 10.1.1.8 0.0.0.3 area 3
Rt99g(config-router)#network 192.168.3.0 0.0.0.255 area 3
Rt99g(config-router)#passive-interface fa0/0
Rt99g(config-router)#end

（8）在路由器 Rt99h 上配置 OSPF

Rt99h(config)#router ospf 404
Rt99h(config-router)#network 10.1.1.12 0.0.0.3 area 4
Rt99h(config-router)#network 192.168.4.0 0.0.0.255 area 4
Rt99h(config-router)#passive-interface fa0/0
Rt99h(config-router)#end

3. 查看 OSPF 各种信息

可以用以下命令查看 OSPF 的各种信息：

Rt99a# show ip route
Rt99a# show ip proto
Rt99a# show ip ospf border-routers
Rt99a# show ip ospf neighbor
Rt99a# show ip ospf database
Rt99a# show ip ospf int se0/1
Rt99a# show ip ospf int se0/0.16
Rt99a# show ip ospf int se0/0.20

以下只列出了在 Rt99b 上查看的结果，请自行查看其他路由器上的信息。

（1）在 Rt99b 上查看路由信息

Rt99b#show ip route
Gateway of last resort is not set
10.0.0.0/8 is variably subnetted, 5 subnets, 2 masks
O IA 10.1.1.0/30 [110/128] via 200.1.1.17, 00:15:37, Serial0/0.16
C 10.1.1.4/30 is directly connected, Serial0/1
O IA 10.1.1.8/30 [110/192] via 200.1.1.17, 00:11:58, Serial0/0.16
[110/192] via 200.1.1.30, 00:11:58, Serial0/0.28
O IA 10.1.1.12/30 [110/128] via 200.1.1.30, 00:11:38, Serial0/0.28

```
C 10.1.1.22/32 is directly connected, Loopback0
O IA 192.168.1.0/24 [110/129] via 200.1.1.17, 00:09:57, Serial0/0.16
O 192.168.2.0/24 [110/65] via 10.1.1.6, 00:08:19, Serial0/1
O IA 192.168.3.0/24 [110/193] via 200.1.1.17, 00:06:23, Serial0/0.16
   [110/193] via 200.1.1.30, 00:06:23, Serial0/0.28
O IA 192.168.4.0/24 [110/129] via 200.1.1.30, 00:04:37, Serial0/0.28
   200.1.1.0/30 is subnetted, 4 subnets
C 200.1.1.16 is directly connected, Serial0/0.16
O 200.1.1.20 [110/128] via 200.1.1.17, 00:15:37, Serial0/0.16
O 200.1.1.24 [110/128] via 200.1.1.30, 00:11:58, Serial0/0.28
C 200.1.1.28 is directly connected, Serial0/0.28
```

（2）在 Rt99b 上查看协议信息

```
Rt99b#show ip proto
Routing Protocol is "ospf 200"
……
```

（3）在 Rt99b 上查看 ABR 信息和 OSPF 邻居信息

```
Rt99b#show ip ospf border-routers
OSPF Process 200 internal Routing Table
……
Rt99b#show ip ospf neighbor
Neighbor ID Pri State Dead Time Address Interface
……
```

（4）在 Rt99b 上查看 OSPF 链路状态数据库

```
Rt99b#show ip ospf database
OSPF Router with ID (10.1.1.22) (Process ID 200)
Router Link States (Area 0)
……
Summary Net Link States (Area 0)
……
Router Link States (Area 2)
……
Summary Net Link States (Area 2)
……
```

（5）在路由器 Rt99b 上显示接口 Se0/1 的 OSPF 配置信息

```
Rt99b#show ip ospf int se0/1
Serial0/1 is up, line protocol is up
Internet address is 10.1.1.5/30, Area 2
Process ID 200, Router ID 10.1.1.22, Network Type POINT-TO-POINT, Cost: 64
```

Transmit Delay is 1 sec, State POINT-TO-POINT, Priority 0
……

（6）在路由器 Rt99b 上显示接口 Se0/0.16 的 OSPF 配置信息

Rt99b#show ip ospf int se0/0.16
Serial0/0.16 is up, line protocol is up
Internet address is 200.1.1.18/30, Area 0
Process ID 200, Router ID 10.1.1.22, Network Type POINT-TO-POINT, Cost: 64
Transmit Delay is 1 sec, State POINT-TO-POINT, Priority 0
……

（7）在路由器 Rt99b 上显示接口 Se0/0.28 的 OSPF 配置信息

Rt99b#show ip ospf int se0/0.28
Serial0/0.28 is up, line protocol is up
Internet address is 200.1.1.29/30, Area 0
Process ID 200, Router ID 10.1.1.22, Network Type POINT-TO-POINT, Cost: 64
Transmit Delay is 1 sec, State POINT-TO-POINT, Priority 0
……

4. 完成网络调试验收

网络配置都以网络上各设备之间能互相通信为目标，因此，在实际工作中，完成网络配置后都要进行通信测试，以验证配置是否正确。如果还有问题，则要进行调试，直到网上的设备全部都能连通。

（1）在路由器 Rt99a 上测试网络连通性

Rt99a#ping 192.168.2.254
.!!!!
Rt99a#ping 192.168.3.254
.!!!!
Rt99a#ping 192.168.4.254
.!!!!

（2）在路由器 Rt99d 上测试网络连通性

Rt99d#ping 192.168.1.254
.!!!!
Rt99d#ping 192.168.2.254
!!!!!
Rt99d#ping 192.168.3.254
!!!!!

（3）在路由器 Rt99b 上查看路由追踪信息

Rt99b#traceroute 192.168.2.254
Type escape sequence to abort.

```
Tracing the route to 192.168.2.254
1 10.1.1.6 0 msec 1 msec 0 msec
2 192.168.2.254 1 msec 0 msec 1 msec

Rt99b#traceroute 192.168.3.254
Type escape sequence to abort.
Tracing the route to 192.168.3.254
1 200.1.1.17 1 msec 1 msec 2 msec
2 200.1.1.25 6 msec 15 msec 3 msec
3 10.1.1.10 4 msec 10 msec 1 msec
4 192.168.3.254 1 msec 4 msec 4 msec

Rt99b#traceroute 192.168.4.254
Type escape sequence to abort.
Tracing the route to 192.168.4.254
1 200.1.1.30 12 msec 4 msec 1 msec
2 10.1.1.14 1 msec 4 msec 1 msec
3 192.168.4.254 2 msec 2 msec 0 msec
```

（4）在 PC99a 上显示路由追踪信息（如图 11-2 所示）

图 11-2　在 PC99a 上显示路由追踪信息

（5）在 PC99c 上显示路由追踪信息（如图 11-3 所示）

（6）保存配置

通过全面测试，确认路由器的配置正确后一定要保存配置，否则路由器重启后又要重新配置和调试。

为了养成良好的习惯，每次完成模拟配置和调试后，都要保存每台路由器的配置，保存配置的命令为 copy run start。在模拟实训时，最后还要保存网络拓扑。

项目 11　OSPF 综合配置

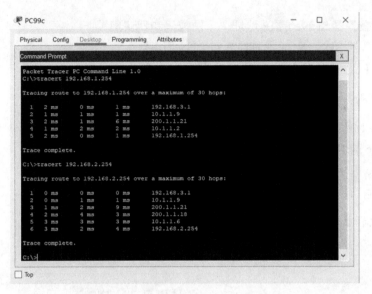

图 11-3　在 PC99c 上显示路由追踪信息

单击【保存】按钮保存相关配置，再将其另存为文件名为 P11002.pkt 的文件备用，然后继续进行实训工作。

5. 模拟上网浏览

在网络中增加一台服务器便可模拟上网浏览。增加服务器之后的网络拓扑如图 11-4 所示，为此，要做如下配置。

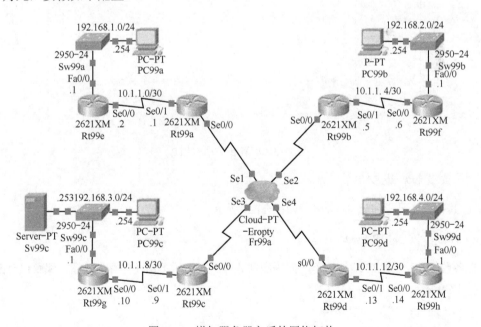

图 11-4　增加服务器之后的网络拓扑

131

- 配置服务器 Sv99c 的 IP 地址 / 子网掩码和默认网关；
- 在服务器 Sv99c 上配置 HTTP 和 DNS 服务；
- 配置 PC 的 IP 地址。

（1）配置服务器的 IP 地址

在服务器上配置 IP 地址、子网掩码和默认网关，然后单击【保存】按钮，保存已有配置。

- Sv99c 的 IP 地址：192.168.3.253；
- Sv99c 的子网掩码：255.255.255.0；
- Sv99c 的默认网关：192.168.3.1。

（2）配置和启动 HTTP

在 Sv99c 上配置和启动 HTTP，如图 11-5 所示。

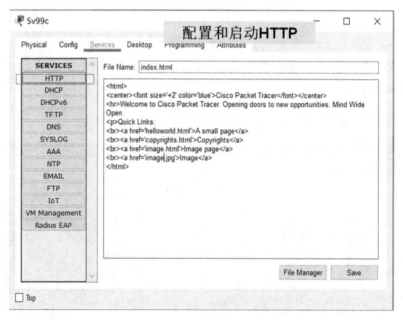

图 11-5　配置和启动 HTTP

（3）配置和启动 DNS

在 Sv99c 上配置和启动 DNS，如图 11-6 所示。

（4）配置 PC 的 IP 地址

在各台 PC 上配置 IP 地址、子网掩码和默认网关，然后单击【保存】按钮，保存已有配置。

① PC99a。

- IP 地址：192.168.1.254；
- 子网掩码：255.255.255.0；
- 默认网关：192.168.1.1；

- DNS 服务器 IP 地址：192.168.3.253。

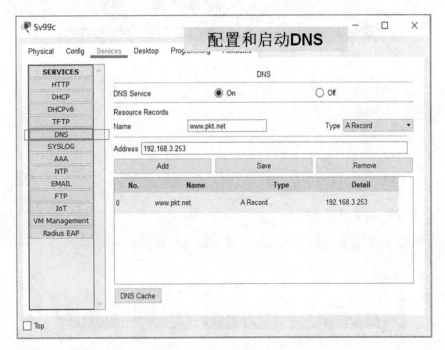

图 11-6　配置和启动 DNS

② PC99b。
- IP 地址：192.168.2.254；
- 子网掩码：255.255.255.0；
- 默认网关：192.168.2.1；
- DNS 服务器 IP 地址：192.168.3.253。

③ PC99c。
- IP 地址：192.168.3.254；
- 子网掩码：255.255.255.0；
- 默认网关：192.168.3.1；
- DNS 服务器 IP 地址：192.168.3.253。

④ PC99d。
- IP 地址：192.168.4.254；
- 子网掩码：255.255.255.0；
- 默认网关：192.168.4.1；
- DNS 服务器 IP 地址：192.168.3.253。

（5）在 PC99a 上上网浏览

在 PC99a 上浏览 http://www.pkt.net，访问成功！如图 11-7 所示。

图 11-7　在 PC99a 上上网浏览

（6）在 PC99b 上上网浏览

在 PC99b 上浏览 http://192.168.3.253，访问成功！如图 11-8 所示。

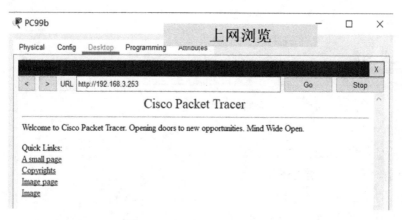

图 11-8　在 PC99b 上上网浏览

（7）在 PC99c 上上网浏览

在 PC99c 上浏览 http://www.pkt.net/helloworld.html，访问成功！如图 11-9 所示。

图 11-9　在 PC99c 上上网浏览

（8）在 PC99d 上上网浏览

在 PC99d 上浏览 http://www.pkt.net/image.html，访问成功！如图 11-10 所示。

项目 11　OSPF 综合配置

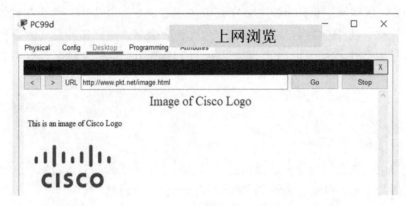

图 11-10　在 PC99d 上上网浏览

至此，上网浏览测试全部完成，功能完全实现。记得退出前保存相关配置！

在一个采用 OSPF 的大型网络中，SPF 算法的反复计算，庞大的路由表和拓扑表的维护以及 LSA（链路状态通告）泛洪等都会占用路由器的资源，因而会降低路由器的运行效率。

OSPF 可利用区域的概念来减小这些不利影响。因为在一个区域内，路由器不需要了解它们所在区域外的拓扑细节。

OSPF 多区域的拓扑结构有如下的优势：

① 降低 SPF 计算频率。
② 减小路由表。
③ 降低了 LSA 的开销。
④ 将网络的不稳定性限制在特定的区域内。

当一个 AS 被划分成几个 OSPF 区域时，根据一台路由器在相应区域之内的作用，可以将路由器分为以下几类。

① 内部路由器：所有直连链路都处于同一个区域的路由器。
② 主干路由器：具有连接区域 0 接口的路由器。
③ 区域边界路由器（ABR）：至少与两个区域相连的路由器。
④ 自治系统边界路由器（ASBR）：与 AS 外部的路由器相连并互相交换路由信息的路由器。

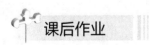

完成上面的模拟实训，将实训过程的截图按顺序粘贴到一个 Word 文件里并用适当的

文字说明你对它的理解；总结本次实训所需要的主要命令及其作用，作为实训报告上交。

实训报告一律以"ID 姓名项目号.doc"为文件名命名，网络拓扑及其配置也以"ID 姓名项目号.pkt"为文件名保存并上交。例如，张三的 ID 为 03，他的文件名为"03 张三 11.doc"和"03 张三 11.pkt"。

思考题

到现在为止，我们学习了哪些路由协议？它们是如何工作的？涉及哪些知识？

请查阅相关信息，对路由协议相关知识和技术做一个总结。

项目 12　静态路由最佳配置

项目描述

若你是某单位的网络管理员,因网络结构简单决定采用静态路由,但为了提高网络效率,要尽量减少路由条目,你必须恰当地配置默认路由和静态汇总路由。

网络拓扑

网络拓扑如图 12-1 所示。

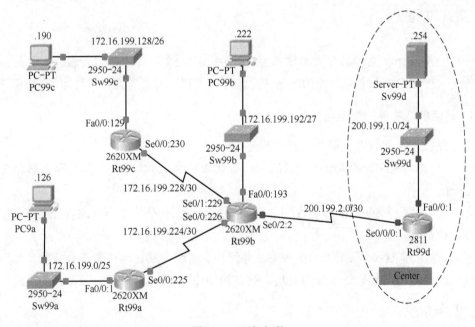

图 12-1　网络拓扑

学习目标

- 熟练掌握静态路由的配置;

- 重点掌握静态汇总路由和默认路由的配置；
- 掌握调试网络的方法。

实训任务分解

① 最佳静态路由的配置与查验。
② 服务器等的配置与测试。

知识点介绍

本项目的实训任务主要是演示项目 09 最佳静态路由的配置，然后增加一些服务器和 PC，为下一个项目做好准备工作。本项目实训在项目 09 的基础上进行，实训环境如图 12-1 所示。项目 09 已完成了基本配置和接口配置，但还未完成 EIGRP 配置，现在要配置静态路由并给出最佳答案。

实训过程

先将 P09002.pkt 文件另存为文件名为 P12001.pkt 的文件备用，然后进行相关配置。
首先，查看拓扑图中路由器的路由信息，可以看到只有直连路由，具体步骤如下所述。

1. 最佳静态路由的配置与查验

通过仔细思考和分析，最佳配置方案是：
① 在路由器 Rt99a 和 Rt99c 上只配置一条默认路由，因为它们与其他路由器相连的路径是唯一的。
② 在路由器 Rt99b 上配置到内部子网的静态路由和到外网的默认路由，因为它与外网的连接路径是唯一的。
③ 在路由器 Rt99d 只配置一条静态汇总路由，而配置默认路由是不合适的，因为它模拟因特网，一般还与其他路由器相连接，只是图中没画而已。

（1）在 Rt99a 上查看路由信息

```
Rt99a# show ip route
Gateway of last resort is not set
172.16.0.0/16 is variably subnetted, 2 subnets, 2 masks
C   172.16.199.0/25 is directly connected, FastEthernet0/0
C   172.16.199.224/30 is directly connected, Serial0/0/0
```

项目 12　静态路由最佳配置

（2）在 Rt99b 上查看路由信息

```
Rt99b# show ip route
Gateway of last resort is not set
     172.16.0.0/16 is variably subnetted, 3 subnets, 2 masks
C    172.16.199.192/27 is directly connected, FastEthernet0/0
C    172.16.199.224/30 is directly connected, Serial0/0
C    172.16.199.228/30 is directly connected, Serial0/1
     200.199.2.0/24 is variably subnetted, 2 subnets, 2 masks
C    200.199.2.0/30 is directly connected, Serial0/2
C    200.199.2.1/32 is directly connected, Serial0/2
```

（3）在 Rt99c 上查看路由信息

```
Rt99c# show ip route
Gateway of last resort is not set
     172.16.0.0/16 is variably subnetted, 2 subnets, 2 masks
C    172.16.199.128/26 is directly connected, FastEthernet0/0
C    172.16.199.228/30 is directly connected, Serial0/0/0
```

（4）在 Rt99d 上查看路由信息

```
Rt99d# show ip route
Gateway of last resort is not set
C    200.199.1.0/24 is directly connected, FastEthernet0/0
     200.199.2.0/24 is variably subnetted, 2 subnets, 2 masks
C    200.199.2.0/30 is directly connected, Serial0/0/0
C    200.199.2.2/32 is directly connected, Serial0/0/0
```

在配置前查看，发现各路由器都只有直连路由。

（5）在 Rt99a 上配置默认路由

```
Rt99a(config)# ip route 0.0.0.0 0.0.0.0 172.16.199.226    //配置默认路由
Rt99a(config)# end
```

（6）在 Rt99b 上配置静态路由和默认路由

```
Rt99b(config)# ip route 172.16.199.0 255.255.255.128 172.16.199.225
Rt99b(config)# ip route 172.16.199.128 255.255.255.192 172.16.199.230
Rt99b(config)# ip route 0.0.0.0 0.0.0.0 200.199.2.1      //配置默认路由
Rt99b(config)# end
```

（7）在 Rt99c 上配置默认路由

```
Rt99c(config)# ip route 0.0.0.0 0.0.0.0 172.16.199.229            //配置默认路由
Rt99c(config)# end
```

（8）在 Rt99d 上配置静态汇总路由

```
Rt99d(config)# ip route 172.16.199.0 255.255.255.0 200.199.2.2   //配置静态汇总路由
```

Rt99d(config)# end

(9) 再次在 Rt99a 上查看路由信息

```
Rt99a# show ip route
Gateway of last resort is 172.16.199.226 to network 0.0.0.0
172.16.0.0/16 is variably subnetted, 2 subnets, 2 masks
C  172.16.199.0/25 is directly connected, FastEthernet0/0
C  172.16.199.224/30 is directly connected, Serial0/0/0
S* 0.0.0.0/0 [1/0] via 172.16.199.226            //默认路由
```

(10) 再次在 Rt99b 上查看路由信息

```
Rt99b# show ip route
Gateway of last resort is 200.199.2.1 to network 0.0.0.0
172.16.0.0/16 is variably subnetted, 5 subnets, 4 masks
S  172.16.199.0/25 [1/0] via 172.16.199.225       //静态路由
S  172.16.199.128/26 [1/0] via 172.16.199.230     //静态路由
C  172.16.199.192/27 is directly connected, FastEthernet0/0
C  172.16.199.224/30 is directly connected, Serial0/0
C  172.16.199.228/30 is directly connected, Serial0/1
200.199.2.0/24 is variably subnetted, 2 subnets, 2 masks
C  200.199.2.0/30 is directly connected, Serial0/2
C  200.199.2.1/32 is directly connected, Serial0/2
S* 0.0.0.0/0 [1/0] via 200.199.2.1               //默认路由
```

(11) 再次在 Rt99c 上查看路由信息

```
Rt99c# show ip route
Gateway of last resort is 172.16.199.229 to network 0.0.0.0
172.16.0.0/16 is variably subnetted, 2 subnets, 2 masks
C  172.16.199.128/26 is directly connected, FastEthernet0/0
C  172.16.199.228/30 is directly connected, Serial0/0/0
S* 0.0.0.0/0 [1/0] via 172.16.199.229            //默认路由
```

(12) 再次在 Rt99d 上查看路由信息

```
Rt99d# show ip route
Gateway of last resort is not set
172.16.0.0/24 is subnetted, 1 subnets
S  172.16.199.0 [1/0] via 200.199.2.2            //静态汇总路由
C  200.199.1.0/24 is directly connected, FastEthernet0/0
200.199.2.0/24 is variably subnetted, 2 subnets, 2 masks
C  200.199.2.0/30 is directly connected, Serial0/0/0
C  200.199.2.2/32 is directly connected, Serial0/0/0
```

在完成路由配置后再次查看，发现各路由器都有了所配置的路由，而且除直连路由外总共只有 6 条静态路由（含默认路由），路由条目已减至最少。

项目 12 静态路由最佳配置

（13）保存配置

```
Rt99a# write
Rt99b# write
Rt99c# write
Rt99d# write
```

单击【保存】按钮保存相关配置到 P12001.pkt 文件中备用，然后再将其另存为文件名为 P12002.pkt 的文件。

2. 服务器等的配置与测试

增加一些服务器和 PC，或是将 PC 改成服务器，为下一个项目做好准备工作。增加服务器和 PC 后的网络拓扑如图 12-2 所示。为各主机配置好 IP 地址、DNS 服务器地址，为各服务器配置好相关服务（如 HTTP、DNS、EMAIL）后，可以做一些有趣的测试。

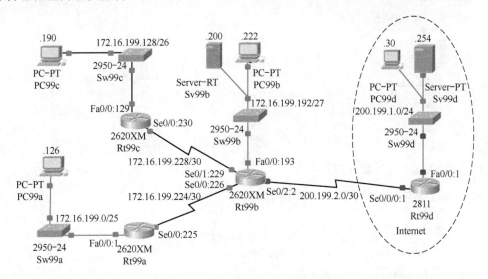

图 12-2 增加服务器和 PC 后的网络拓扑

（1）配置各主机

在各主机上，按下表中的数据配置 IP 地址、子网掩码、默认网关和 DNS 服务器地址。

主 机 名	IP 地 址	子网掩码	默 认 网 关	DNS 服务器地址
Sv99b	172.16.199.200	255.255.255.224	172.16.199.193	172.16.199.200
PC99a	172.16.199.126	255.255.255.128	172.16.199.1	172.16.199.200
PC99b	172.16.199.222	255.255.255.224	172.16.199.193	172.16.199.200
PC99c	172.16.199.190	255.255.255.192	172.16.199.129	172.16.199.200
PC99d	200.199.1.30	255.255.255.0	200.199.1.1	200.199.1.254
Sv99d	200.199.1.254	255.255.255.0	200.199.1.1	200.199.1.254

(2) 在 Sv99b 上配置 HTTP 服务

在 Sv99b 上配置 HTTP 服务，如图 12-3 所示。

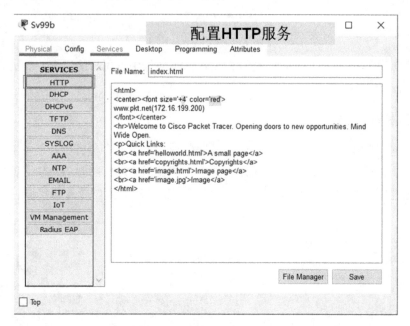

图 12-3　在 Sv99b 上配置 HTTP 服务

(3) 在 Sv99d 上配置 HTTP 服务

在 Sv99d 上配置 HTTP 服务，如图 12-4 所示。

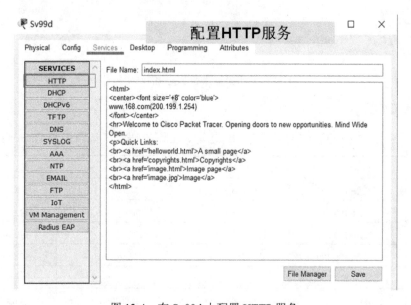

图 12-4　在 Sv99d 上配置 HTTP 服务

（4）在 Sv99b 上配置 DNS 服务

在 Sv99b 上配置 DNS 服务，如图 12-5 所示。

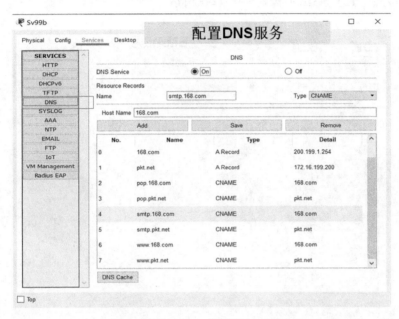

图 12-5　在 Sv99b 上配置 DNS 服务

（5）在 Sv99d 上配置 DNS 服务

在 Sv99d 上配置 DNS 服务，如图 12-6 所示。

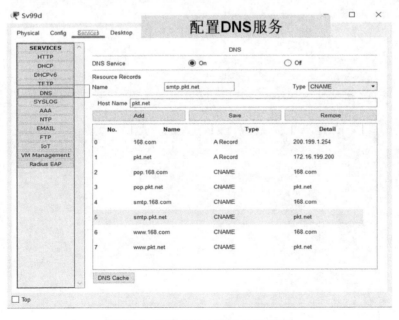

图 12-6　在 Sv99d 上配置 DNS 服务

（6）在 Sv99b 上配置 EMAIL 服务

在 Sv99b 上配置 EMAIL 服务，如图 12-7 所示。

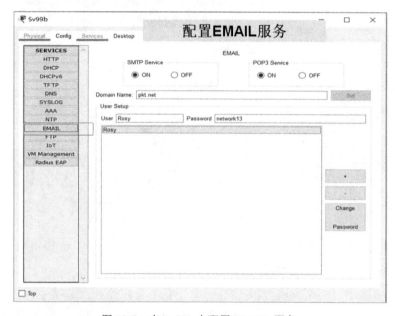

图 12-7　在 Sv99b 上配置 EMAIL 服务

（7）在 Sv99d 上配置 EMAIL 服务

在 Sv99d 上配置 EMAIL 服务，如图 12-8 所示。

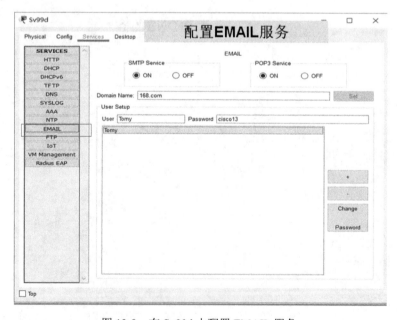

图 12-8　在 Sv99d 上配置 EMAIL 服务

项目 12　静态路由最佳配置

（8）在 PC99a 上浏览 http://www.pkt.net

如图 12-9 所示，在 PC99a 上浏览 http://www.pkt.net。访问成功！

图 12-9　在 PC99a 上浏览 http://www.pkt.net

（9）在 PC99c 上浏览 http://www.168.com

如图 12-10 所示，在 PC99c 上浏览 http://www.168.com。访问成功！

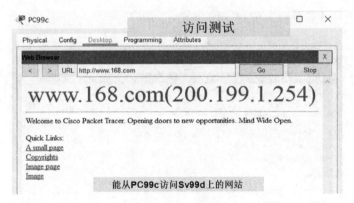

图 12-10　在 PC99c 上浏览 http://www.168.com

（10）在 Sv99d 上浏览 http://www.pkt.net

如图 12-11 所示，在 Sv99d 上浏览 http://www.pkt.net。访问成功！

图 12-11　在 Sv99d 上浏览 http://www.pkt.net

（11）在 PC99b 上完成邮件收发配置

在 PC99b 上完成邮件收发配置，如图 12-12 所示。

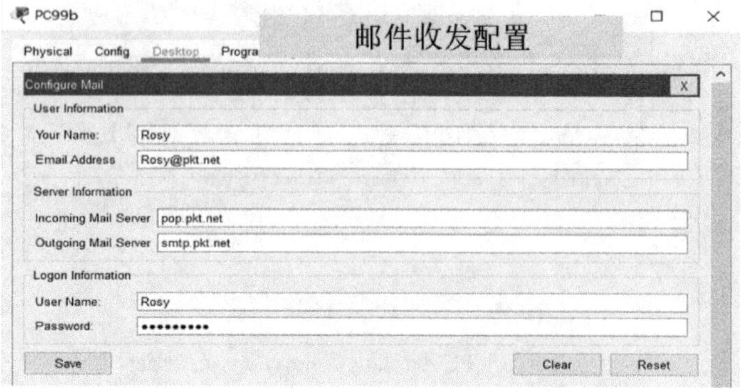

图 12-12　在 PC99b 上完成邮件收发配置

（12）在 PC99d 上完成邮件收发配置

在 PC99d 上完成邮件收发配置，如图 12-13 所示。

图 12-13　在 PC99d 上完成邮件收发配置

（13）在 PC99d 上编写邮件并发送

如图 12-14 所示，在 PC99d 上编写邮件并发送。发送成功！

图 12-14　在 PC99d 上编写邮件并发送

项目 12　静态路由最佳配置

（14）在 PC99b 上接收并查看邮件

如图 12-15 所示，在 PC99b 上接收并查看邮件。接收成功！

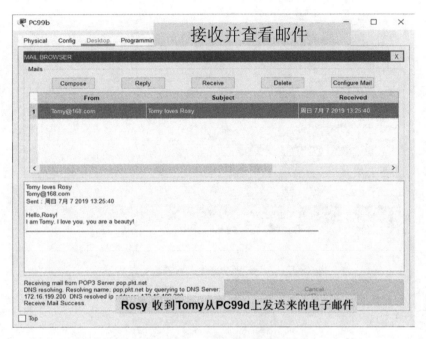

图 12-15　在 PC99b 上接收并查看邮件

（15）在 PC99b 上编写并发送邮件

如图 12-16 所示，在 PC99b 上编写并发送邮件。发送成功！

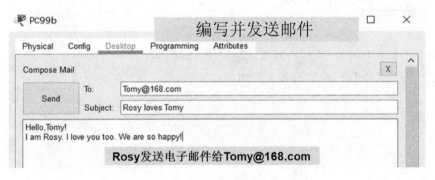

图 12-16　在 PC99b 上编写并发送邮件

（16）在 PC99d 上接收并查看邮件

如图 12-17 所示，在 PC99d 上接收并查看邮件。接收成功！
至此，上网浏览以及邮件发送与接收等全部测试完成，功能完全实现。
最后单击【保存】按钮，保存相关配置到文件 P12002.pkt 中。

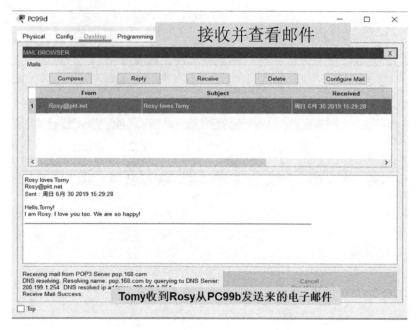

图 12-17　在 PC99d 上接收并查看邮件

学习总结

本项目为项目 13 做好了准备工作，也为项目 09 的课后作业提供了一个标准答案。我们不仅要熟练掌握静态路由的配置方法，还必须认真思考如何配置才是最佳的。

思考题

已有一个配置私有 IP 地址的网络，如何接入因特网？
在下一个项目中我们将学习有关的知识和技术。

项目 13 NAT 及其配置

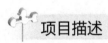

某企业构建了一个内部网络（简称内网，已配置好私有 IP 地址），现要把它与因特网连接，要求内网的 1 台服务器能对外提供 HTTP 和 EMAIL 等服务，其他机器只访问因特网、不提供服务。申请获得的合法 IP 地址块为 201.8.199.0/29。

若你是该企业的网络管理员，你怎么做？

已有一个配置私有 IP 地址的网络，现在要把它接入因特网，怎么实现？

利用 NAT（网络地址转换）技术可以很容易实现上述需求，而且还能极大地节约注册 IP 地址，并对因特网用户隐藏内部网络的 IP 地址，提高网络安全性。

下面我们将重点学习 NAT 技术的应用。

网络拓扑如图 13-1 所示。

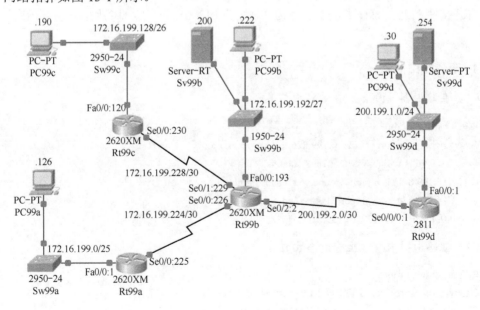

图 13-1 网络拓扑

149

学习目标

- 重点掌握静态 NAT、动态 NAT 和 PAT 的配置；
- 了解 NAT 的作用；
- 掌握 HTTP、EMAIL、DNS 服务器的配置；
- 掌握检验 NAT 的方法；
- 掌握调试网络的方法。

实训任务分解

① 查看路由信息。
② 配置静态 NAT、动态 NAT 和 PAT 功能。
③ 配置 HTTP、EMAIL、DNS 服务器和 EMAIL 客户端。
④ 测试和检验 NAT、HTTP、EMAIL、DNS、FTP 配置。

实训过程

本项目实训在项目 12 的基础上进行，实训环境如图 13-1 所示。其中，只是增加了服务器和 PC，静态路由的配置已在项目 12 中完成。根据需要，后面将只修改 Rt99d 的静态路由；最重要的数据是所获的 IP 地址块 201.8.199.0/29。

打开文件 P12002.pkt，再将其另存为文件名为 P13001.pkt 的文件备用，然后开始进行实训工作。

1. 查看路由信息

（1）在路由器 Rt99a 上查看路由信息

```
Rt99a# show ip route
Gateway of last resort is 172.16.199.226 to network 0.0.0.0
172.16.0.0/16 is variably subnetted, 2 subnets, 2 masks
C 172.16.199.0/25 is directly connected, FastEthernet0/0
C 172.16.199.224/30 is directly connected, Serial0/0
S* 0.0.0.0/0 [1/0] via 172.16.199.226                //默认路由
```

（2）在路由器 Rt99b 上查看路由信息

```
Rt99b# show ip route
Gateway of last resort is 200.199.2.1 to network 0.0.0.0
172.16.0.0/16 is variably subnetted, 5 subnets, 4 masks
```

```
S 172.16.199.0/25 [1/0] via 172.16.199.225          //静态路由
S 172.16.199.128/26 [1/0] via 172.16.199.230        //静态路由
C 172.16.199.192/27 is directly connected, FastEthernet0/0
C 172.16.199.224/30 is directly connected, Serial0/0
C 172.16.199.228/30 is directly connected, Serial0/1
    200.199.2.0/24 is variably subnetted, 2 subnets, 2 masks
C 200.199.2.0/30 is directly connected, Serial0/2
C 200.199.2.1/32 is directly connected, Serial0/2
S* 0.0.0.0/0 [1/0] via 200.199.2.1                  //默认路由
```

（3）在路由器 Rt99c 上查看路由信息

```
Rt99c# show ip route
Gateway of last resort is 172.16.199.229 to network 0.0.0.0
    172.16.0.0/16 is variably subnetted, 2 subnets, 2 masks
C 172.16.199.128/26 is directly connected, FastEthernet0/0
C 172.16.199.228/30 is directly connected, Serial0/0
S* 0.0.0.0/0 [1/0] via 172.16.199.229               //默认路由
```

（4）查在路由器 Rt99d 上查看路由信息

```
Rt99d# show ip route
Gateway of last resort is not set
    172.16.0.0/24 is subnetted, 1 subnets
S 172.16.199.0 [1/0] via 200.199.2.2                //静态汇总路由
C 200.199.1.0/24 is directly connected, FastEthernet0/0
    200.199.2.0/24 is variably subnetted, 2 subnets, 2 masks
C 200.199.2.0/30 is directly connected, Serial0/0/0
C 200.199.2.2/32 is directly connected, Serial0/0/0
```

2. 配置静态 NAT、动态 NAT 和 PAT 功能

NAT 有三种类型：静态 NAT、动态 NAT 和 PAT。

静态 NAT：将内部 IP 地址与合法 IP 地址进行一对一的转换。对外提供 HTTP、EMAIL、FTP 等服务的服务器的 IP 地址只能采用静态 NAT 方式转换。

动态 NAT：采用动态分配的方式将合法 IP 地址池中的地址一对一地映射为内部 IP 地址。

PAT（端口地址转换）：把内部 IP 地址映射到配有合法 IP 地址的不同端口上，从而可实现多对一的映射。PAT 对于节省合法 IP 地址是最为有效的。

本实训项目将学习配置每种类型的 NAT 功能，但都只在图 13-1 中的 Rt99b 上完成。

（1）在路由器 Rt99b 上配置 NAT 功能

```
Rt99b(config)# ip nat inside source static 172.16.199.200 201.8.199.1   //配置静态 NAT 功能
Rt99b(config)# ip nat pool NP1 201.8.199.2 201.8.199.2 netmask 255.255.255.248
```

Rt99b(config)# ip nat pool NP2 201.8.199.3 201.8.199.6 netmask 255.255.255.248
Rt99b(config)# access-list 1 permit 172.16.199.0 0.0.0.127
Rt99b(config)# access-list 1 permit 172.16.199.128 0.0.0.63
Rt99b(config)# access-list 2 permit 172.16.199.192 0.0.0.31
Rt99b(config)# ip nat inside source list 2 pool NP2   //配置动态 NAT 功能
Rt99b(config)# ip nat inside source list 1 pool NP1 overload //配置 PAT 功能
//PAT 与动态 NAT 配置命令的区别是多了一个 overload 参数

对上面配置命令的补充解释如下：

Rt99b(config)# ip nat inside source static 172.16.199.200 201.8.199.1
//配置静态 NAT 功能，以满足 Sv99b 对外提供服务的需求
Rt99b(config)# ip nat pool NP1 201.8.199.2 201.8.199.2 netmask 255.255.255.248
//定义合法 IP 地址池 NP1，用于配置 PAT 功能
//请注意 PAT 只用了一个合法 IP 地址 201.8.199.2/29
Rt99b(config)# ip nat pool NP2 201.8.199.3 201.8.199.6 netmask 255.255.255.248
//定义合法 IP 地址池 NP2，用于配置动态 NAT 功能
Rt99b(config)# access-list 1 permit 172.16.199.0 0.0.0.127
Rt99b(config)# access-list 1 permit 172.16.199.128 0.0.0.63
//定义允许进行 PAT 的内部 IP 地址范围
//注意命令最后的参数为反掩码，它等于 255.255.255.255−子网掩码
Rt99b(config)# access-list 2 permit 172.16.199.192 0.0.0.31
//定义允许进行动态 NAT 的内部 IP 地址范围
Rt99b(config)# ip nat inside source list 1 pool NP1 overload //配置 PAT 功能
Rt99b(config)# ip nat inside source list 2 pool NP2   //配置动态 NAT 功能

（2）在路由器 Rt99b 上配置 NAT 的内部与外部接口

Rt99b(config)# int fa0/0
Rt99b(config-if)# ip nat inside   //定义 Fa0/0 为 NAT 内部接口
Rt99b(config-if)# int se0/0
Rt99b(config-if)# ip nat inside   //定义 Se0/0 为 NAT 内部接口
Rt99b(config-if)# int se0/1
Rt99b(config-if)# ip nat inside   //定义 Se0/1 为 NAT 内部接口
Rt99b(config-if)# int se0/2
Rt99b(config-if)# ip nat outside   //定义 Se0/2 为 NAT 外部接口
Rt99b(config-if)# end

请注意：NAT 功能的完整配置是由（1）和（2）两步完成的。

（3）在路由器 Rt99d 上配置静态路由

Rt99d(config)# ip route 201.8.199.0 255.255.255.248 200.199.2.2   //增加静态路由
Rt99d(config)# end
Rt99d# show ip route static
172.16.0.0/24 is subnetted, 1 subnets

项目 13　NAT 及其配置

```
S    172.16.199.0 [1/0] via 200.199.2.2                //此时应删除该路由
     201.8.199.0/29 is subnetted, 1 subnets
S    201.8.199.0 [1/0] via 200.199.2.2
```
注意：外网一般不知道也不应该知道内网的 IP 地址。

（4）在路由器 Rt99d 上删除静态路由

```
Rt99d(config)# no ip route 172.16.199.0 255.255.255.0 200.199.2.2   //删除静态路由
Rt99d(config)# end
Rt99d# show ip route static
     201.8.199.0/29 is subnetted, 1 subnets
S    201.8.199.0 [1/0] via 200.199.2.2                //唯一到本企业网的静态路由
//这样外网就不能直接用内网 IP 地址访问内网了
```

（5）在路由器 Rt99b 上查看 NAT 相关信息

```
Rt99b# show ip nat translation
Pro Inside global Inside local Outside local Outside global
---    201.8.199.1   172.16.199.200   ---   ---
```

（6）保存配置

```
Rt99b# copy run start
Rt99d# write
```

单击模拟器的【保存】按钮保存配置到 P13001.pkt 文件中备用，然后再将其另存为文件名为 P13002.pkt 的文件。

3. 配置 HTTP、EMAIL、DNS 服务器和 EMAIL 客户端

为了访问因特网，各 PC 和服务器都要增配所用 DNS 服务器的 IP 地址。

在 Sv99b 上配置 HTTP、DNS、EMAIL、FTP 服务，网站域名为 www.pkt.net；邮箱地址为 Rosy@pkt.net，邮箱密码为 net2013；该服务器给企业网提供 DNS 服务，但不为外网提供 DNS 服务。

在 Sv99d 上配置 DNS、HTTP、EMAIL 服务，网站域名为 www.168.com；邮箱地址为 Tomy@168.com，邮箱密码为 cisco13；该服务器为因特网提供 DNS 服务，但不为企业网提供 DNS 服务。

上述任务只要在项目 12 的基础上做少量修改便可完成。

（1）在服务器 Sv99d 上修改 DNS 配置

在服务器 Sv99d 上修改 DNS 配置，如图 13-2 所示。

DNS 的配置很重要，它提供如下域名的解析服务：

- HTTP 服务中配置的网站域名；
- EMAIL 服务中配置的域名，电子邮件接收服务器的域名，电子邮件发送服务器的域名。

如果 DNS 配置不当，则电子邮件的收发和基于域名的 Web 浏览等都会出现问题。

图 13-2　在服务器 Sv99d 上修改 DNS 配置

（2）在服务器 Sv99b 上增设 EMAIL 账户

在服务器 Sv99b 上增设 EMAIL 账户，如图 13-3 所示。

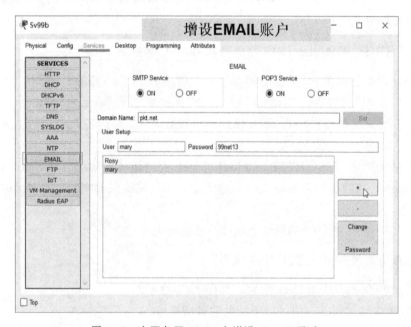

图 13-3　在服务器 Sv99b 上增设 EMAIL 账户

（3）在服务器 Sv99b 上配置 FTP

在服务器 Sv99b 上配置 FTP，如图 13-4 所示。

项目 13　NAT 及其配置

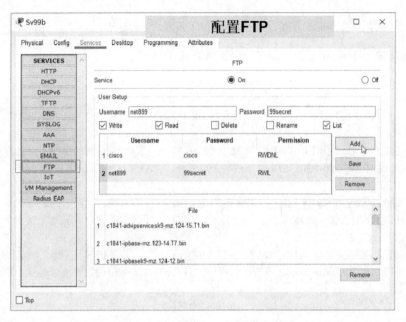

图 13-4　在服务器 Sv99b 上配置 FTP

（4）在 PC99a 上配置 EMAIL 客户端信息

在 PC99a 上配置 EMAIL 客户端信息，如图 13-5 所示。

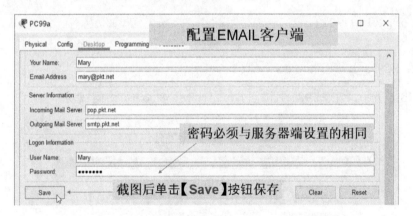

图 13-5　在 PC99a 上配置 EMAIL 客户端信息

4. 测试和检验 NAT、HTTP、EMAIL、DNS、FTP 配置

（1）在路由器 Rt99b 上开启 NAT debug 模式

```
Rt99b# debug ip nat                    //启动 debug 模式
IP NAT debugging is on
```

（2）在路由器 Rt99d 上 ping Sv99b，测试是否可访问

```
Rt99d# ping 201.8.199.1                //ping 服务器 Sv99b
.!!!!
```

注意：外网只能访问具有静态 NAT 功能的主机。

（3）在路由器 Rt99b 上查看静态 NAT 相关信息

```
IP NAT debugging is on
Rt99b#
NAT: s=200.199.2.1, d=201.8.199.1->172.16.199.200 [1]      //请求包的目的地址转换
Rt99b#
NAT: s=200.199.2.1, d=201.8.199.1->172.16.199.200 [2]      //请求包的目的地址转换
NAT*: s=172.16.199.200->201.8.199.1, d=200.199.2.1 [1]     //回应包的源地址转换
Rt99b#
NAT: s=200.199.2.1, d=201.8.199.1->172.16.199.200 [3]      //请求包的目的地址转换
NAT*: s=172.16.199.200->201.8.199.1, d=200.199.2.1 [2]     //回应包的源地址转换
Rt99b#
NAT: s=200.199.2.1, d=201.8.199.1->172.16.199.200 [4]      //进来时目的地址转换
NAT*: s=172.16.199.200->201.8.199.1, d=200.199.2.1 [3]     //出去时源地址转换
Rt99b#
NAT: s=200.199.2.1, d=201.8.199.1->172.16.199.200 [5]      //进来时目的地址转换
```

以上结果显示，静态 NAT 功能起作用。

（4）在路由器 Rt99d 上继续测试服务器

```
Rt99d# ping 201.8.199.6                                    //不能访问
……
Rt99d# ping 201.8.199.2                                    //不能访问
……
```

（5）在 PC99a 上访问 http://www.168.com

如图 13-6 所示，在 PC99a 上访问 http://www.168.com。访问成功！

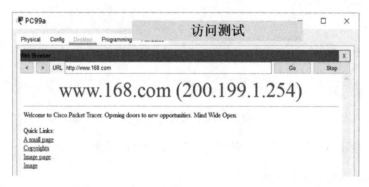

图 13-6　在 PC99a 上访问 http://www.168.com

（6）在路由器 Rt99b 上查看 PAT 相关信息

```
Rt99b#
NAT: s=172.16.199.126->201.8.199.2, d=200.199.1.254 [2]    //出去时源地址转换
Rt99b#
```

项目 13　NAT 及其配置

```
NAT*: s=172.16.199.126->201.8.199.2, d=200.199.1.254 [3]        //出去时源地址转换
NAT*: s=200.199.1.254, d=201.8.199.2->172.16.199.126 [1]        //进来时目的地址转换
NAT*: s=172.16.199.126->201.8.199.2, d=200.199.1.254 [4]        //出去时源地址转换
NAT*: s=172.16.199.126->201.8.199.2, d=200.199.1.254 [5]        //出去时源地址转换
NAT*: s=200.199.1.254, d=201.8.199.2->172.16.199.126 [2]        //进来时目的地址转换
NAT*: s=172.16.199.126->201.8.199.2, d=200.199.1.254 [6]        //出去时源地址转换
Rt99b#
NAT*: s=200.199.1.254, d=201.8.199.2->172.16.199.126 [3]        //进来时目的地址转换
NAT*: s=172.16.199.126->201.8.199.2, d=200.199.1.254 [8]        //出去时源地址转换
NAT: s=172.16.199.126->201.8.199.2, d=200.199.1.254 [9]         //出去时源地址转换
```

以上结果显示，PAT 功能起作用。

（7）在 PC99c 上访问 http://www.pkt.net

如图 13-7 所示，在 PC99c 上访问 http://www.pkt.net。访问成功！

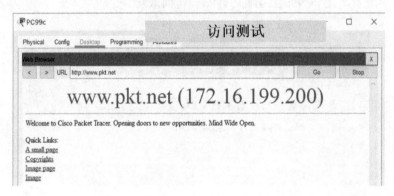

图 13-7　在 PC99c 上访问 http://www.pkt.net

可是在路由器 Rt99b 上未查看到 NAT 相关信息。

以上结果表明，本次访问属于内网间访问，不需要 NAT 功能。

（8）在 PC99b 上访问 http://www.168.com

如图 13-8 所示，在 PC99b 上访问 http://www.168.com。访问成功！

图 13-8　在 PC99b 上访问 http://www.168.com

（9）在路由器 Rt99b 上查看动态 NAT 相关信息

```
Rt99b#
NAT: s=172.16.199.222->201.8.199.3, d=200.199.1.254 [2]            //出去时源地址转换
NAT*: s=200.199.1.254, d=201.8.199.3->172.16.199.222 [10]          //进来时目的地址转换
NAT*: s=172.16.199.222->201.8.199.3, d=200.199.1.254 [3]           //出去时源地址转换
NAT*: s=172.16.199.222->201.8.199.3, d=200.199.1.254 [4]           //出去时源地址转换
NAT*: s=200.199.1.254, d=201.8.199.3->172.16.199.222 [11]          //进来时目的地址转换
NAT*: s=172.16.199.222->201.8.199.3, d=200.199.1.254 [5]           //出去时源地址转换
NAT*: s=200.199.1.254, d=201.8.199.3->172.16.199.222 [12]          //进来时目的地址转换
NAT*: s=172.16.199.222->201.8.199.3, d=200.199.1.254 [6]           //出去时源地址转换
```

以上结果显示，动态 NAT 功能起作用。

（10）在 PC99c 上访问 Sv99d

如图 13-9 所示，在 PC99c 上访问 Sv99d。访问成功！

图 13-9　在 PC99c 上访问 Sv99d

（11）再次在路由器 Rt99b 上查看 PAT 相关信息

```
Rt99b#
NAT: s=172.16.199.190->201.8.199.2, d=200.199.1.254 [15]           //出去时源地址转换
NAT*: s=200.199.1.254, d=201.8.199.2->172.16.199.190 [13]          //进来时目的地址转换
NAT*: s=172.16.199.190->201.8.199.2, d=200.199.1.254 [16]          //出去时源地址转换
NAT*: s=172.16.199.190->201.8.199.2, d=200.199.1.254 [17]          //出去时源地址转换
NAT*: s=200.199.1.254, d=201.8.199.2->172.16.199.190 [14]          //进来时目的地址转换
NAT*: s=172.16.199.190->201.8.199.2, d=200.199.1.254 [18]          //出去时源地址转换
NAT*: s=200.199.1.254, d=201.8.199.2->172.16.199.190 [15]          //进来时目的地址转换
NAT*: s=172.16.199.190->201.8.199.2, d=200.199.1.254 [19]          //出去时源地址转换
```

以上结果显示，PAT 功能起作用。

（12）在 PC99d 上访问 Sv99b

如图 13-10 所示，在 PC99d 上访问 Sv99b。访问成功！

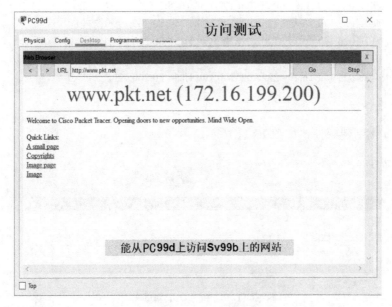

图 13-10　在 PC99d 上访问 Sv99b

（13）再次在路由器 Rt99b 上查看静态 NAT 相关信息

Rt99b#
NAT: s=200.199.1.30, d=201.8.199.1->172.16.199.200 [2]　　//进来时目的地址转换
NAT*: s=172.16.199.200->201.8.199.1, d=200.199.1.30 [25]　　//出去时源地址转换
NAT*: s=200.199.1.30, d=201.8.199.1->172.16.199.200 [3]　　//进来时目的地址转换
NAT*: s=200.199.1.30, d=201.8.199.1->172.16.199.200 [4]　　//进来时目的地址转换
NAT*: s=172.16.199.200->201.8.199.1, d=200.199.1.30 [26]　　//出去时源地址转换
NAT*: s=200.199.1.30, d=201.8.199.1->172.16.199.200 [5]　　//进来时目的地址转换
NAT*: s=172.16.199.200->201.8.199.1, d=200.199.1.30 [27]　　//出去时源地址转换
NAT*: s=200.199.1.30, d=201.8.199.1->172.16.199.200 [6]　　//进来时目的地址转换

以上结果显示，静态 NAT 功能起作用。

（14）在路由器 Rt99b 上关闭 debug 模式并查看 NAT 统计信息

Rt99b# no debug ip nat　　　　　　　　　　　　　　//关闭 debug 模式
IP NAT debugging is off
Rt99b# show ip nat statistics　　　　　　　　　　　　//查看 NAT 统计信息
Total translations: 7 (1 static, 6 dynamic, 6 extended)
……

（15）在路由器 Rt99b 上查看 NAT 统计信息

Rt99b# show ip nat translations

```
Pro Inside global Inside local Outside local Outside global
--- 201.8.199.1 172.16.199.200 --- ---
tcp 201.8.199.1:1025 172.16.199.200:1025 200.199.1.254:80 200.199.1.254:80
tcp 201.8.199.1:80 172.16.199.200:80 200.199.1.30:1025 200.199.1.30:1025
tcp 201.8.199.2:1025 172.16.199.126:1025 200.199.1.254:80 200.199.1.254:80
tcp 201.8.199.2:1026 172.16.199.126:1026 200.199.1.254:80 200.199.1.254:80
tcp 201.8.199.2:1027 172.16.199.190:1027 200.199.1.254:80 200.199.1.254:80
tcp 201.8.199.3:1025 172.16.199.222:1025 200.199.1.254:80 200.199.1.254:80
```

（16）新建邮件

在 PC99a 上新建邮件，如图 13-11 所示。

图 13-11 在 PC99a 上新建邮件

（17）编辑并发送邮件

编辑好邮件后单击【Send】按钮发送邮件，如图 13-12 所示。

图 13-12 编辑好邮件后单击【Send】按钮发送邮件

（18）在 PC99d 上查看收到的邮件

在 PC99d 上查看到来自 PC99a 的邮件，如图 13-13 所示，单击【Receive】按钮接收邮件。

（19）在 PC99d 上发送邮件

在 PC99d 上发送邮件，如图 13-14 所示。

（20）在 PC99a 上查看从 PC99d 发送来的邮件

在 PC99a 上查看从 PC99d 发送来的邮件，如图 13-15 所示，单击【Receive】按钮接收邮件。

项目 13　NAT 及其配置

图 13-13　在 PC99d 上查看到来自 PC99a 的邮件

图 13-14　在 PC99d 上发送邮件

图 13-15　在 PC99a 上查看从 PC99d 发送来的邮件

（21）再次在路由器 Rt99b 上查看 NAT 统计信息

```
Rt99b# show ip nat translations
Pro Inside global Inside local Outside local Outside global
--- 201.8.199.1 172.16.199.200 --- ---
tcp 201.8.199.1:1025 172.16.199.200:1025 200.199.1.254:80 200.199.1.254:80
tcp 201.8.199.1:1026 172.16.199.200:1026 200.199.1.254:25 200.199.1.254:25
tcp 201.8.199.1:25 172.16.199.200:25 200.199.1.254:1025 200.199.1.254:1025
tcp 201.8.199.1:80 172.16.199.200:80 200.199.1.30:1025 200.199.1.30:1025
tcp 201.8.199.2:1025 172.16.199.126:1025 200.199.1.254:80 200.199.1.254:80
tcp 201.8.199.2:1026 172.16.199.126:1026 200.199.1.254:80 200.199.1.254:80
tcp 201.8.199.2:1027 172.16.199.190:1027 200.199.1.254:80 200.199.1.254:80
tcp 201.8.199.3:1025 172.16.199.222:1025 200.199.1.254:80 200.199.1.254:80
```

（22）再次查看 NAT 统计信息

```
Rt99b# show ip nat statistics
Total translations: 9(1 static, 8 dynamic, 8 extended)
……
```

（23）在 PC99c 上登录 FTP 服务器并上传文件

```
C:\> dir                                          //查看本地文件
Volume in drive C has no label.
Volume Serial Number is 5E12-4AF3
Directory of C:\
2/7/2106 14:28 PM 26 sampleFile.txt
26 bytes 1 File(s)
C:\> ftp 172.16.199.200                           //登录 FTP 服务器
Trying to connect...172.16.199.200
Connected to 172.16.199.200
220- Welcome to PT Ftp server
Username:net899
331- Username ok, need password
Password:                                         //输入 99secret
230- Logged in
(passive mode On)
ftp> put sampleFile.txt                           //上传文件
Writing file sampleFile.txt to 172.16.199.200:
File transfer in progress...
[Transfer complete - 26 bytes]
26 bytes copied in 0.016 secs (1625 bytes/sec)
ftp>
```

（24）在服务器 Sv99b 上查看 PC99c 上传的文件

在服务器 Sv99b 上查看 PC99c 上传的文件，如图 13-16 所示。

项目13 NAT及其配置

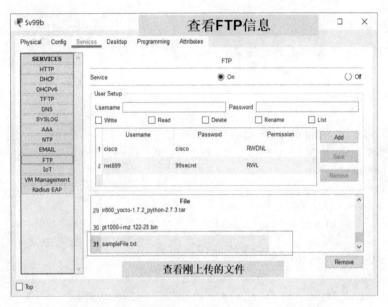

图13-16 在服务器Sv99b上查看PC99c上传的文件

（25）在PC99d上登录FTP服务器

```
Packet Tracer PC Command Line 1.0
C:\> dir                                          //查看本地文件
Volume in drive C has no label.
Volume Serial Number is 5E12-4AF3
Directory of C:\
2/7/2106 14:28 PM 26 sampleFile.txt
26 bytes 1 File(s)
C:\> ftp 201.9.199.1                              //登录FTP服务器
Trying to connect...201.8.199.1
Connected to 201.8.199.1
220- Welcome to PT Ftp server
Username:cisco
331- Username ok , need password
Password:Cisco
230- Logged in
(Passive mode on)
```

（26）查看FTP服务器上的文件列表

```
ftp> dir
Listing /ftp directory from 201.8.199.1:
0 : FileA.txt 26
1 : asa842-k8.bin 5571584
……
33 : pt3000-i6q4l2-mz.121-22.EA4.bin 3117390
```

163

34 : sampleFile.txt 26

（27）下载 FTP 服务器上的文件

```
ftp> rename sampleFile.txt FileA.txt        //重新命名 FTP 服务器上的文件
Renaming sampleFile.txt
ftp>
[OK Renamed file successfully from sampleFile.txt to FileA.txt]
ftp> get FileA.txt                          //下载 FTP 服务器上的文件
Reading file FileA.txt from 201.8.199.1:
File transfer in progress...
[Transfer complete - 26 bytes]
26 bytes copied in 0.01 secs (2600 bytes/sec)
ftp> dir                                    //查看 FTP 服务器上的文件
Listing /ftp directory from 201.8.199.1:
0 : FileA.txt 26
1 : asa842-k8.bin 5571584
……
33 : pt3000-i6q4l2-mz.121-22.EA4.bin 3117390
34 : sampleFile.txt 26
```

（28）查看下载的本地文件

```
ftp> help                                   //查看 FTP 命令
……
ftp> quit                                   //退出 FTP
221- Service closing control connection.
C:\> dir                                    //查看本地文件
Volume in drive C has no label.
Volume Serial Number is 5E12-4AF3
Directory of C:\
1/1/1970 8:0 PM 26 FileA.txt                //刚下载的文件
2/7/2106 14:28 PM 26 sampleFile.txt
52 bytes 2 File(s)
```

完成以上操作后单击【保存】按钮，将配置保存到 P13002.pkt 文件中备用，再将其另存为文件名为 P14001.pkt 的文件，为下一个项目做好准备工作。退出时选择保存。

学习总结

因特网中合法 IP 地址短缺已成为一个十分突出的问题。NAT（Network Address Translation）是解决 IP 地址短缺的重要手段。NAT 是一个 IETF 标准。

NAT 技术使得一个私有网络可以通过注册 IP 地址连接到外部网络，内部网络中具有 NAT 功能的路由器在发送数据包之前，负责把内部 IP 地址转换成外部合法 IP 地址。通过应用 NAT 技术可将每个局域网节点的 IP 地址转换成一个合法 IP 地址，反之亦然。它也可

项目 13 NAT 及其配置

以应用到防火墙技术中，把个别 IP 地址隐藏起来不被外界发现，对内部网络设备起到保护作用，同时，它还可以使网络超越地址的限制，合理地使用网络中的公有 IP 地址和私有 IP 地址。

NAT 有三种类型：静态 NAT、动态 NAT 和 PAT。

静态 NAT 将内部 IP 地址与合法 IP 地址进行一对一的转换。对外提供 HTTP、EMAIL、FTP 等服务的服务器的 IP 地址只能采用静态 NAT 方式转换。

动态 NAT 采用动态分配的方法将合法 IP 地址池中的地址一对一地映射为内部 IP 地址。

PAT（端口地址转换）则把内部 IP 地址映射到配有合法 IP 地址的不同端口上，从而可实现多对一的映射。PAT 对于节省合法 IP 地址非常有效。

静态 NAT 和 PAT 最实用，动态 NAT 很少用。

课后作业

完成上面的模拟实训，将实训过程的截图按顺序粘贴到一个 Word 文件里并用适当的文字说明你对它的理解；总结本次实训所需要的主要命令及其作用，作为实训报告上交。

实训报告一律以"ID 姓名项目号.doc"为文件名命名，网络拓扑及其配置也以"ID 姓名项目号.pkt"为文件名保存并上交。例如，张三的 ID 为 03，他的文件名为"03 张三 13.doc"和"03 张三 13.pkt"。

思考题

如果将本实训中的动态 NAT 改为 PAT，如何配置？

网络安全是令用户头痛的问题，如何控制用户对网络的访问？如何防范来自因特网的攻击？

在下一个项目中我们将学习有关的知识和技术。

项目 14　ACL 基本配置

项目描述

某企业构建好了一个网络后，出于安全考虑，要求对网络中设备的访问权限进行一些控制，譬如：
- 某个路由器只允许某个主机远程登录；
- 禁止某两个子网的主机互访；
- 禁止子网 A 中的主机访问子网 B，但允许子网 B 中的主机访问子网 A。

若你是该企业的网络管理员，你怎么做？网络安全是令用户头痛的问题，如何控制用户对网络的访问？如何防范来自因特网的攻击呢？

网络拓扑

网络拓扑如图 14-1 所示。

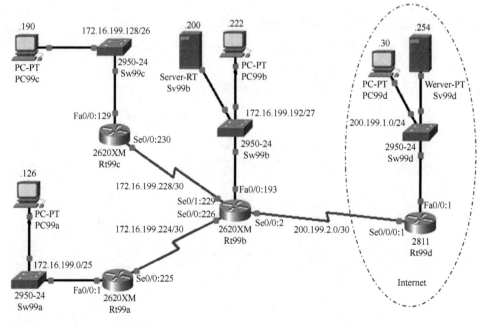

图 14-1　网络拓扑

项目 14　ACL 基本配置

学习目标

- 重点掌握配置 ACL 的基本方法；
- 了解 ACL 的作用；
- 掌握检验 ACL 的方法；
- 掌握调试网络的方法。

实训任务分解

① 实训 1：在路由器 Rt99a 上配置标准 ACL，使它只允许 IP 地址为 172.16.199.190 中的主机远程登录。

② 实训 2：在路由器 Rt99c 上配置标准 ACL，禁止 172.16.199.0/25 子网中的主机访问子网 172.16.199.128/26。

③ 实训 3：将 Rt99c 上配置的标准 ACL 改为扩展 ACL，禁止子网 172.16.199.0/25 中的主机访问 172.16.199.128/26 子网，但允许子网 172.16.199.128/26 中的主机访问子网 172.16.199.0/25。

④ 实训 4：测试验收。

知识点介绍

路由器用于连接不同的网络，处于网络的边界，因此，对转发的数据包进行过滤可以控制用户对网络的访问。一般都在与外网相连的边界路由器上配置包过滤防火墙。包过滤可通过 ACL（访问控制列表）来实现。

实训过程

实训环境和配置数据：本项目在项目 13 的基础上进行，实训环境如图 14-1 所示。静态路由的配置已完成，网络已全部连通。启动模拟器，打开文件 P13002.pkt，再将其另存为文件名为 P14001.pkt 的文件（若已有该文件直接打开即可）。

1. 实训 1：在路由器 Rt99a 上配置标准 ACL

在路由器 Rt99a 上配置标准 ACL，使它只允许 IP 地址为 172.16.199.190 的主机远程登录。标准 ACL 只使用 IP 包中的源 IP 地址对数据包进行过滤，因此最简单，但用途也有限，它的表号范围为[1, 99] 或 [1300, 1999]。

创建标准 ACL 的命令如下：

Rt99a(config)# access-list 4 permit 172.16.199.190 0.0.0.0

在 vty 下应用 ACL 的命令如下：

Rt9a(config-line)# access-class 4 in

（1）在 Rt99a 上配置标准 ACL

Rt99a(config)# access-list 4 permit 172.16.199.190 0.0.0.0   //创建标准 ACL
Rt99a(config)# line vty 0 4
Rt99a(config-line)# access-class 4 in   //在 vty 的入方向应用 ACL
Rt99a(config-line)# password 99vty04
Rt99a(config-line)# login
Rt99a(config-line)# logg sync
Rt99a(config-line)# exec-timeout 0 0
Rt99a(config-line)# end
Rt99a# show access-lists   //显示 ACL
Standard IP access list 4
10 permit host 172.16.199.190

注意：标准 ACL 的表号在[1,99]范围内即可。

（2）在 Rt99b 上远程登录 172.16.199.1，测试 ACL 是否起作用

Rt99b# telnet 172.16.199.1   //远程登录被拒绝
Trying 172.16.199.1 ...Open
[Connection to 172.16.199.1 closed by foreign host]

结果：远程登录被拒绝。

（3）在 PC99c 上远程登录 172.16.199.1，测试 ACL 是否起作用

C:\> telnet 172.16.199.1
Trying 172.16.199.1 ...Open
User Access Verification
Password: //输入口令：99vty04
Rt99a>en
Password: //输入口令：99secret
Rt99a# show access-lists
Standard IP access list 4
    10 permit host 172.16.199.190 (2 match(es))
Rt99a# disable
Rt99a> logout
[Connection to 172.16.199.1 closed by foreign host]
C:\>

结果：远程登录被允许。

（4）在 PC99b 上远程登录 172.16.199.1，测试 ACL 是否起作用

```
C:\> telnet 172.16.199.1                              //远程登录被拒绝
Trying 172.16.199.1 ...Open
[Connection to 172.16.199.1 closed by foreign host]
```
结果：远程登录被拒绝。

（5）显示匹配 ACL 规则的情况

```
Rt99a# show access-lists                              //显示 ACL 信息
Standard IP access list 4
    10 permit host 172.16.199.190 (2 match(es))       //与该规则已匹配的次数
```

（6）保存配置

```
Rt99a# copy run start
```

2. 实训 2：在路由器 Rt99c 上配置标准 ACL

在 Rt99c 上配置标准 ACL，禁止子网 172.16.199.0/25 中的主机访问子网 172.16.199.128/26。由于子网 172.16.199.128/26 中的主机访问子网 172.16.199.0/25 的返回数据包被丢弃，故标准 ACL 实际禁止这两个子网互访。

（1）在 Rt99c 上配置标准 ACL

```
Rt99c(config)# access-list 2 deny 172.16.199.0 0.0.0.127    //创建标准 ACL
Rt99c(config)# access-list 2 permit any                     //在已有 ACL 表项中添加访问控制项
Rt99c(config)# int se0/0
Rt99c(config-if)# ip access-group 2 in                      //在 Se0/0 入方向应用 ACL
Rt99c(config-if)# end
Rt99c# show access-lists                                    //显示 ACL
Standard IP access list 2
    10 deny 172.16.199.0 0.0.0.127
    20 permit any
```
注意：标准 ACL 的表号在[1,99]范围内即可。

（2）在 Rt99b 上启动监控功能

```
Rt99b# debug  ip  packet
Packet debugging is on                                      //启动 IP 数据包转发监控功能
```

（3）在 Rt99c 上启动监控功能

Rt99c# debug　ip　packet
Packet debugging is on   //启动 IP 数据包转发监控功能

（4）在 PC99a 上发起与 PC99c 的通信，测试 ACL 是否起作用（如图 14-2 所示）

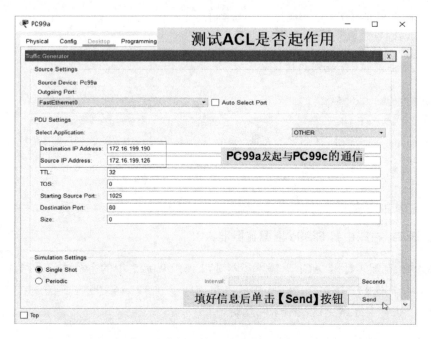

图 14-2　在 PC99a 上发起与 PC99c 的通信，测试 ACL 是否起作用

（5）在 Rt99b 上观察从 172.16.199.126 到 172.16.199.190 的 IP 数据包转发情况

Rt99b#
IP: tableid=0, s=172.16.199.126 (Serial0/0), d=172.16.199.190 (Serial0/1), routed via RIB
IP: s=172.16.199.126 (Serial0/0), d=172.16.199.190 (Serial0/1), g=172.16.199.230, len 28, forward
IP: tableid=0, s=172.16.199.230 (Serial0/1), d=172.16.199.126 (Serial0/0), routed via RIB
IP: s=172.16.199.230 (Serial0/1), d=172.16.199.126 (Serial0/0), g=172.16.199.225, len 56, forward
　　　　　　　　　　　　　　　　　//IP 数据包转发监控

结果：从 Se0/0 接收的一个到 PC99c 的请求数据包被转发到 Se0/1。

（6）在 Rt99c 上观察从 172.16.199.126 到 172.16.199.190 的 IP 数据包转发情况

Rt99c# debug　ip　packet   //启动 IP 数据包转发监控功能
Packet debugging is on
Rt99c#

结果：请求数据包被拦截，没有到达 IP 层。

项目 14 ACL 基本配置

（7）在 PC99b 上发起与 PC99c 的通信，测试 ACL 是否起作用（如图 14-3 所示）

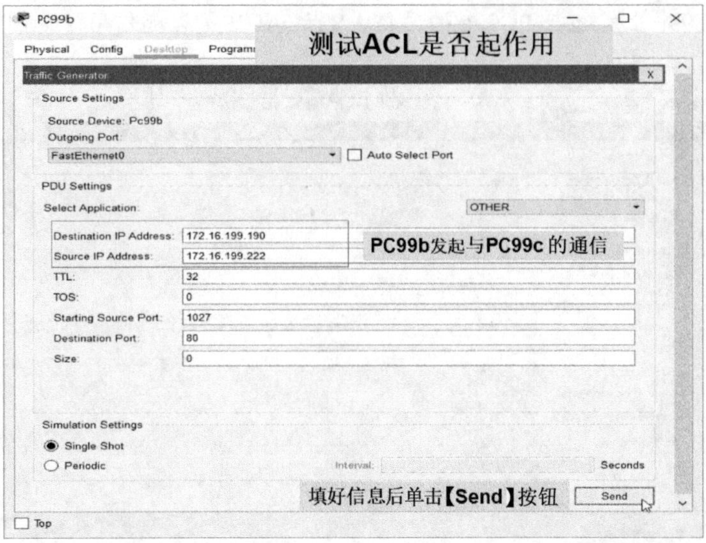

图 14-3 在 PC99b 上发起与 PC99c 的通信，测试 ACL 是否起作用

（8）在 Rt99b 上观察从 172.16.199.222 到 172.16.199.190 的 IP 数据包转发情况

Rt99b# debug ip packet   //启动 IP 数据包转发监控功能
Packet debugging is on
IP: tableid=0, s=172.16.199.222 (FastEthernet0/0), d=172.16.199.190 (Serial0/1), routed via RIB
IP: s=172.16.199.222 (FastEthernet0/0), d=172.16.199.190 (Serial0/1), g=172.16.199.230, len 28, forward
//以上 2 行显示了请求数据包相关信息
IP: tableid=0, s=172.16.199.190 (Serial0/1), d=172.16.199.222 (FastEthernet0/0), routed via RIB
IP: s=172.16.199.190 (Serial0/1), d=172.16.199.222 (FastEthernet0/0), g=172.16.199.222, len 56, forward
//以上 2 行显示了应答数据包相关信息
Rt99b#

结果：请求数据包和应答数据包都被正常转发。

（9）在 Rt99c 上观察从 172.16.199.222 到 172.16.199.190 的 IP 数据包转发情况

Rt99c# debug ip packet   //启动 IP 数据包转发监控功能
Packet debugging is on
Rt99c#
IP: tableid=0, s=172.16.199.222 (Serial0/0), d=172.16.199.190 (FastEthernet0/0), routed via RIB
IP: s=172.16.199.222 (Serial0/0), d=172.16.199.190 (FastEthernet0/0), g=172.16.199.190, len 28, forward
//以上 2 行显示了请求数据包相关信息
IP: tableid=0, s=172.16.199.190 (FastEthernet0/0), d=172.16.199.222 (Serial0/0), routed via RIB
IP: s=172.16.199.190 (FastEthernet0/0), d=172.16.199.222 (Serial0/0), g=172.16.199.229, len 56, forward
//以上 2 行显示了应答数据包相关信息
Rt99c#

结果：请求数据包和应答数据包都被正常转发。

（10）在 PC99c 上发起与 PC99a 的通信，测试 ACL 是否起作用（如图 14-4 所示）

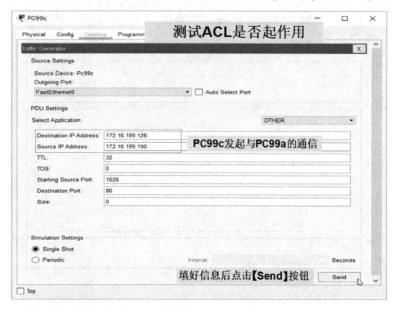

图 14-4 在 PC99c 上发起与 PC99a 的通信，测试 ACL 是否起作用

（11）在 Rt99c 上观察从 172.16.199.190 到 172.16.199.126 的 IP 数据包转发情况

```
Rt99c# debug  ip  packet                    //启动 IP 数据包转发监控功能
Packet debugging is on
Rt99c#
IP: tableid=0, s=172.16.199.190 (FastEthernet0/0), d=172.16.199.126 (Serial0/0), routed via RIB
IP: s=172.16.199.190 (FastEthernet0/0), d=172.16.199.126 (Serial0/0), g=172.16.199.229, len 28, forward
//以上 2 行显示监测到请求数据包
IP: tableid=0, s=172.16.199.230 (local), d=172.16.199.126 (Serial0/0), routed via RIB
IP: s=172.16.199.230 (local), d=172.16.199.126 (Serial0/0), len 84, sending
Rt99c#
```

结果：请求数据包被转发出去，应答数据包被拦截。

（12）在 Rt99b 上观察从 172.16.199.190 到 172.16.199.126 的 IP 数据包转发情况

```
Rt99b# debug  ip  packet                    //启动 IP 数据包转发监控功能
Packet debugging is on
Rt99b#
IP: tableid=0, s=172.16.199.190 (Serial0/1), d=172.16.199.126 (Serial0/0), routed via RIB
IP: s=172.16.199.190 (Serial0/1), d=172.16.199.126 (Serial0/0), g=172.16.199.225, len 28, forward
//以上 2 行显示监测到请求数据包
IP: tableid=0, s=172.16.199.126 (Serial0/0), d=172.16.199.190 (Serial0/1), routed via RIB
IP: s=172.16.199.126 (Serial0/0), d=172.16.199.190 (Serial0/1), g=172.16.199.230, len 56, forward
```

IP: tableid=0, s=172.16.199.230 (Serial0/1), d=172.16.199.126 (Serial0/0), routed via RIB
IP: s=172.16.199.230 (Serial0/1), d=172.16.199.126 (Serial0/0), g=172.16.199.225, len 84, forward
//以上 4 行显示监测到应答数据包

结果：请求数据包和应答数据包都被正常转发。

（13）在 PC99c 上发起与 PC99b 的通信，测试 ACL 是否起作用（如图 14-5 所示）

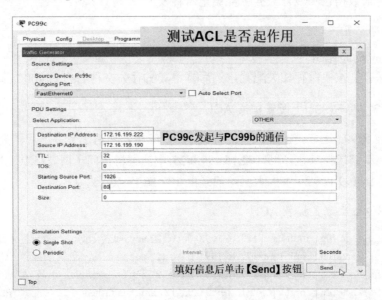

图 14-5 在 PC99c 上发起与 PC99b 的通信，测试 ACL 是否起作用

（14）在 Rt99c 上观察从 172.16.199.190 到 172.16.199.222 的 IP 数据包转发情况

Rt99c# debug ip packet   //启动 IP 数据包转发监控功能
Packet debugging is on
Rt99c#
IP: tableid=0, s=172.16.199.190 (FastEthernet0/0), d=172.16.199.222 (Serial0/0), routed via RIB
IP: s=172.16.199.190 (FastEthernet0/0), d=172.16.199.222 (Serial0/0), g=172.16.199.229, len 28, forward
//以上 2 行显示监测到请求数据包
IP: tableid=0, s=172.16.199.222 (Serial0/0), d=172.16.199.190 (FastEthernet0/0), routed via RIB
IP: s=172.16.199.222 (Serial0/0), d=172.16.199.190 (FastEthernet0/0), g=172.16.199.190, len 56, forward
//以上 2 行显示监测到应答数据包
Rt99c#

结果：请求数据包和应答数据包都被正常转发。

（15）在 Rt99b 上观察从 172.16.199.190 到 172.16.199.222 的 IP 数据包转发情况

Rt99b# debug ip packet   //启动 IP 数据包转发监控功能
Packet debugging is on
Rt99b#
IP: tableid=0, s=172.16.199.190 (Serial0/1), d=172.16.199.222 (FastEthernet0/0), routed via RIB
IP: s=172.16.199.190 (Serial0/1), d=172.16.199.222 (FastEthernet0/0), g=172.16.199.222, len 28, forward

//以上 2 行显示监测到请求数据包
IP: tableid=0, s=172.16.199.222 (FastEthernet0/0), d=172.16.199.190 (Serial0/1), routed via RIB
IP: s=172.16.199.222 (FastEthernet0/0), d=172.16.199.190 (Serial0/1), g=172.16.199.230, len 56, forward
//以上 2 行显示监测到应答数据包
Rt99b#

结果：请求数据包和应答数据包都被正常转发。

（16）保存配置

Rt99c# copy   run   start

单击【保存】按钮保存相关配置，然后再将其另存为文件名为 P14002.pkt 的文件。

3. 实训 3：在 Rt99c 上配置扩展 ACL

在路由器 Rt99c 上取消上面配置的标准 ACL 应用，然后配置并应用扩展 ACL，以实现以下访问控制需求：禁止子网 172.16.199.0/25 中的主机访问子网 172.16.199.128/26，但允许子网 172.16.199.128/26 中的主机访问子网 172.16.199.0/25。

（1）在 Rt99c 上配置扩展 ACL

```
Rt99c(config)# acc 106 permit icmp 172.16.199.0 0.0.0.127 any echo-reply       //配置 ACL
Rt99c(config)# acc 106 permit icmp 172.16.199.0 0.0.0.127 any ttl-exceeded     //配置 ACL
Rt99c(config)# acc 106 permit icmp 172.16.199.0 0.0.0.127 any unreachable      //配置 ACL
Rt99c(config)# acc 106 permit tcp 172.16.199.0 0.0.0.127 any established       //配置 ACL
Rt99c(config)# acc 106 deny ip 172.16.199.0 0.0.0.127 any                      //配置 ACL
Rt99c(config)# acc 106 permit ip any any                                       //配置 ACL
Rt99c(config)# int se0/0
Rt99c(config-if)# ip access-group 106   in                                     //应用 ACL
Rt99c(config-if)# end
```

注意：扩展 ACL 的表号在[100,199]范围内即可。

（2）在 Rt99c 上启动监控功能

```
Rt99c# show access-list                                //显示 ACL 信息
Standard IP access list 2
10 deny 172.16.199.0 0.0.0.127 (3 match(es))
20 permit any (5 match(es))
Extended IP access list 106
10 permit icmp 172.16.199.0 0.0.0.127 any echo-reply
20 permit icmp 172.16.199.0 0.0.0.127 any ttl-exceeded
30 permit icmp 172.16.199.0 0.0.0.127 any unreachable
40 permit tcp 172.16.199.0 0.0.0.127 any established
50 deny ip 172.16.199.0 0.0.0.127 any
60 permit ip any any
Rt99c# debug   ip   packet                             //启动 IP 数据包转发监控功能
Packet debugging is on
```

项目 14　ACL 基本配置

（3）在 Rt99b 启动监控功能

Rt99b# debug　ip　packet　　　　　　　　　　　//启动 IP 数据包转发监控功能
Packet debugging is on

（4）在 PC99a 上发起与 PC99c 的通信，测试 ACL 是否起作用（如图 14-6 所示）

图 14-6　在 PC99a 上发起与 PC99c 的通信，测试 ACL 是否起作用

（5）在 Rt99b 上观察从 172.16.199.126 到 172.16.199.190 的 IP 数据包转发情况

Rt99b# debug　ip　packet　　　　　　　　　　　//启动 IP 数据包转发监控功能
Packet debugging is on
IP: tableid=0, s=172.16.199.126 (Serial0/0), d=172.16.199.190 (Serial0/1), routed via RIB
IP: s=172.16.199.126 (Serial0/0), d=172.16.199.190 (Serial0/1), g=172.16.199.230, len 28, forward
//以上 2 行显示了请求数据包相关信息

IP: tableid=0, s=172.16.199.230 (Serial0/1), d=172.16.199.126 (Serial0/0), routed via RIB
IP: s=172.16.199.230 (Serial0/1), d=172.16.199.126 (Serial0/0), g=172.16.199.225, len 56, forward
//以上 2 行显示了应答数据包相关信息

结果：请求数据包和应答数据包都被正常转发。

（6）在 Rt99c 上观察从 172.16.199.126 到 172.16.199.190 的 IP 数据包转发情况

Rt99c# debug　ip　packet
Packet debugging is on

结果：请求数据包被拦截，没有到达 IP 层。

（7）在 PC99c 上发起与 PC99a 的通信，测试 ACL 是否起作用（如图 14-7 所示）

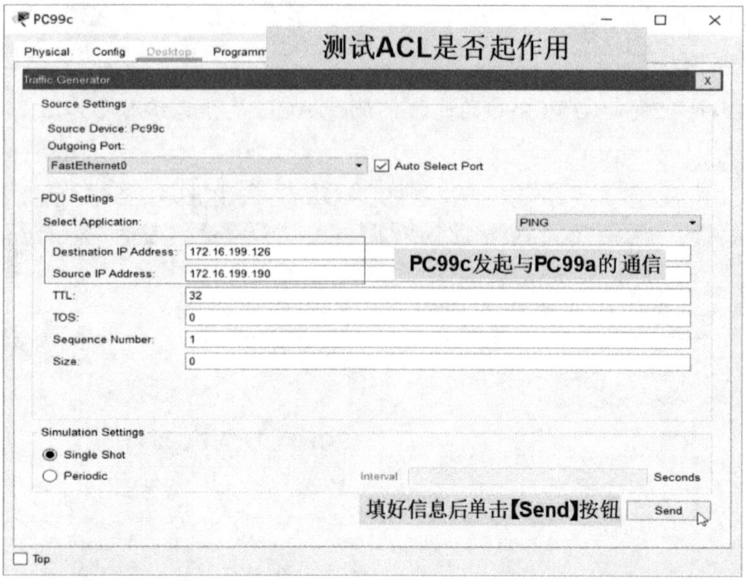

图 14-7 在 PC99c 上发起与 PC99a 的通信，测试 ACL 是否起作用

（8）在 Rt99b 上观察从 172.16.199.190 到 172.16.199.126 的 IP 数据包转发情况

```
Rt99b# debug ip packet                    //启动 IP 数据包转发监控功能
Packet debugging is on
IP: tableid=0, s=172.16.199.190 (Serial0/1), d=172.16.199.126 (Serial0/0), routed via RIB
IP: s=172.16.199.190 (Serial0/1), d=172.16.199.126 (Serial0/0), g=172.16.199.225, len 28, forward
//以上 2 行显示了请求数据包相关信息
IP: tableid=0, s=172.16.199.126 (Serial0/0), d=172.16.199.190 (Serial0/1), routed via RIB
IP: s=172.16.199.126 (Serial0/0), d=172.16.199.190 (Serial0/1), g=172.16.199.230, len 28, forward
//以上 2 行显示了应答数据包相关信息
```

结果：请求数据包和应答数据包都被正常转发。

（9）在 Rt99c 上观察从 172.16.199.190 到 172.16.199.126 的 IP 数据包转发情况

```
Rt99c# debug ip packet                    //启动 IP 数据包转发监控功能
Packet debugging is on
Rt99c#
IP: tableid=0, s=172.16.199.190 (FastEthernet0/0), d=172.16.199.126 (Serial0/0), routed via RIB
IP: s=172.16.199.190 (FastEthernet0/0), d=172.16.199.126 (Serial0/0), g=172.16.199.229, len 28, forward
//以上 2 行显示了请求数据包相关信息
IP: tableid=0, s=172.16.199.126 (Serial0/0), d=172.16.199.190 (FastEthernet0/0), routed via RIB
IP: s=172.16.199.126 (Serial0/0), d=172.16.199.190 (FastEthernet0/0), g=172.16.199.190, len 28, forward
//以上 2 行显示了应答数据包相关信息
```

结果：请求数据包和应答数据包都被正常转发。

（10）在 PC99b 上发起与 PC99c 的通信，测试 ACL 是否起作用（如图 14-8 所示）

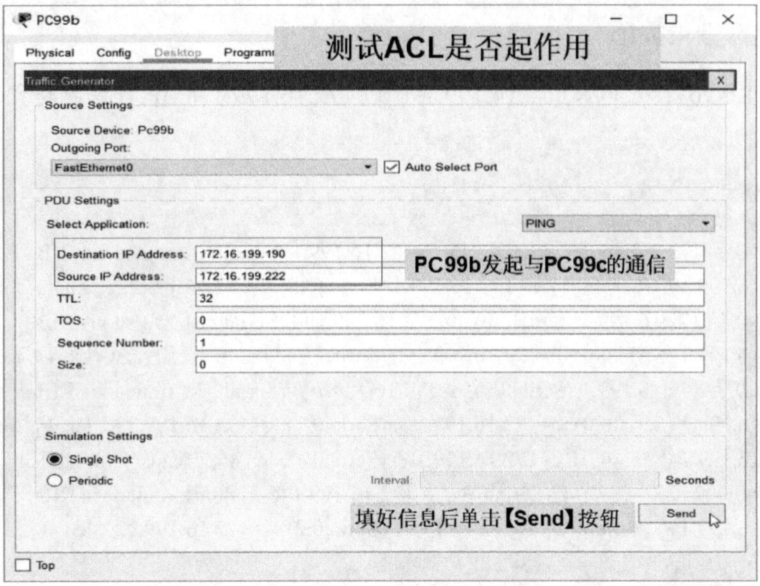

图 14-8　在 PC99b 上发起与 PC99c 的通信，测试 ACL 是否起作用

（11）在 Rt99b 上观察从 172.16.199.222 到 172.16.199.190 的 IP 数据包转发情况

Rt99b# debug ip packet
Packet debugging is on
Rt99b#
IP: tableid=0, s=172.16.199.222 (FastEthernet0/0), d=172.16.199.190 (Serial0/1), routed via RIB
IP: s=172.16.199.222 (FastEthernet0/0), d=172.16.199.190 (Serial0/1), g=172.16.199.230, len 28, forward
IP: tableid=0, s=172.16.199.190 (Serial0/1), d=172.16.199.222 (FastEthernet0/0), routed via RIB
IP: s=172.16.199.190 (Serial0/1), d=172.16.199.222 (FastEthernet0/0), g=172.16.199.222, len 28, forward
结果：请求数据包和应答数据包都被正常转发。

（12）在 Rt99c 上观察从 172.16.199.222 到 172.16.199.190 的 IP 数据包转发情况

Rt99c# debug ip packet
Packet debugging is on
Rt99c#
IP: tableid=0, s=172.16.199.222 (Serial0/0), d=172.16.199.190 (FastEthernet0/0), routed via RIB
IP: s=172.16.199.222 (Serial0/0), d=172.16.199.190 (FastEthernet0/0), g=172.16.199.190, len 28, forward
IP: tableid=0, s=172.16.199.190 (FastEthernet0/0), d=172.16.199.222 (Serial0/0), routed via RIB
IP: s=172.16.199.190 (FastEthernet0/0), d=172.16.199.222 (Serial0/0), g=172.16.199.229, len 28, forward
结果：请求数据包和应答数据包都被正常转发。

（13）在 PC99c 上远程登录 172.16.199.1，测试 ACL 是否起作用

C:\> telnet 172.16.199.1　　　　　　　　　　　　//远程登录被允许
Trying 172.16.199.1 ...Open

User Access Verification
Password: //输入口令 99vty04
结果：远程登录被允许。

（14）在 Rt99b 上观察从 172.16.199.190 到 172.16.199.1 的 IP 数据包转发情况

Rt99b# debug ip packet
Packet debugging is on
Rt99b#
IP: tableid=0, s=172.16.199.190 (Serial0/1), d=172.16.199.1 (Serial0/0), routed via RIB
IP: s=172.16.199.190 (Serial0/1), d=172.16.199.1 (Serial0/0), g=172.16.199.225, len 41, forward
IP: tableid=0, s=172.16.199.1 (Serial0/0), d=172.16.199.190 (Serial0/1), routed via RIB
IP: s=172.16.199.1 (Serial0/0), d=172.16.199.190 (Serial0/1), g=172.16.199.230, len 41, forward
IP: tableid=0, s=172.16.199.1 (Serial0/0), d=172.16.199.190 (Serial0/1), routed via RIB
IP: s=172.16.199.1 (Serial0/0), d=172.16.199.190 (Serial0/1), g=172.16.199.230, len 46, forward
IP: s=172.16.199.1 (Serial0/0), d=172.16.199.190 (Serial0/1), len 46, encapsulation failed
IP: tableid=0, s=172.16.199.190 (Serial0/1), d=172.16.199.1 (Serial0/0), routed via RIB
IP: s=172.16.199.190 (Serial0/1), d=172.16.199.1 (Serial0/0), g=172.16.199.225, len 40, forward
Rt99b#

结果：请求数据包和应答数据包都被正常转发。

（15）在 Rt99c 上观察从 172.16.199.190 到 172.16.199.1 的 IP 数据包转发情况

Rt99c# debug ip packet
Packet debugging is on
Rt99c#
IP: tableid=0, s=172.16.199.190 (FastEthernet0/0), d=172.16.199.1 (Serial0/0), routed via RIB
IP: s=172.16.199.190 (FastEthernet0/0), d=172.16.199.1 (Serial0/0), g=172.16.199.229, len 41, forward
IP: tableid=0, s=172.16.199.1 (Serial0/0), d=172.16.199.190 (FastEthernet0/0), routed via RIB
IP: s=172.16.199.1 (Serial0/0), d=172.16.199.190 (FastEthernet0/0), g=172.16.199.190, len 41, forward
IP: tableid=0, s=172.16.199.1 (Serial0/0), d=172.16.199.190 (FastEthernet0/0), routed via RIB
IP: s=172.16.199.1 (Serial0/0), d=172.16.199.190 (FastEthernet0/0), g=172.16.199.190, len 46, forward
IP: s=172.16.199.1 (Serial0/0), d=172.16.199.190 (FastEthernet0/0), len 46, encapsulation failed
IP: tableid=0, s=172.16.199.190 (FastEthernet0/0), d=172.16.199.1 (Serial0/0), routed via RIB
IP: s=172.16.199.190 (FastEthernet0/0), d=172.16.199.1 (Serial0/0), g=172.16.199.229, len 40, forward
Rt99c#

结果：请求数据包和应答数据包都被正常转发。

（16）在 Rt99c 上显示 ACL 的匹配情况

Rt99c# show access-lists   //显示 ACL 信息
Standard IP access list 2
10 deny 172.16.199.0 0.0.0.127 (3 match(es))
20 permit any (5 match(es))
Extended IP access list 106
10 permit icmp 172.16.199.0 0.0.0.127 any echo-reply (2 match(es))
20 permit icmp 172.16.199.0 0.0.0.127 any ttl-exceeded

```
30 permit icmp 172.16.199.0 0.0.0.127 any unreachable
40 permit tcp 172.16.199.0 0.0.0.127 any established (113 match(es))
50 deny ip 172.16.199.0 0.0.0.127 any (2 match(es))
60 permit ip any any (3 match(es))
```

(17)在 Rt99c 上清零 ACL 计数器

```
Rt99c# clear access-list counters                //清零 ACL 计数器
Rt99c# show access-lists                         //显示 ACL 信息
Standard IP access list 2
10 deny 172.16.199.0 0.0.0.127
20 permit any
Extended IP access list 106
10 permit icmp 172.16.199.0 0.0.0.127 any echo-reply
20 permit icmp 172.16.199.0 0.0.0.127 any ttl-exceeded
30 permit icmp 172.16.199.0 0.0.0.127 any unreachable
40 permit tcp 172.16.199.0 0.0.0.127 any established
50 deny ip 172.16.199.0 0.0.0.127 any
60 permit ip any any
```

(18)在 Rt99c 上配置命名 ACL

```
Rt99c(config)# ip access-list extended in-filter            //配置命名 ACL
Rt99c(config-ext-nacl)# permit icmp 172.16.199.0 0.0.0.127 any echo-reply
Rt99c(config-ext-nacl)# permit icmp 172.16.199.0 0.0.0.127 any ttl-exceeded
Rt99c(config-ext-nacl)# permit icmp 172.16.199.0 0.0.0.127 any unreachable
Rt99c(config-ext-nacl)# permit tcp 172.16.199.0 0.0.0.127 any established
Rt99c(config-ext-nacl)# deny ip 172.16.199.0 0.0.0.127 any
Rt99c(config-ext-nacl)# permit ip any any
//以上 6 行：配置命名 ACL
Rt99c(config-ext-nacl)# exit
Rt99c(config)# int se0/0
Rt99c(config-if)# ip access-group in-filter in    //应用命名为 in-filter 的 ACL，覆盖编号为 106 的 ACL
Rt99c(config-if)# end
```

(19)在 Rt99c 上显示已配置的 ACL 信息

```
Rt99c# show access-lists
Standard IP access list 2
10 deny 172.16.199.0 0.0.0.127
20 permit any
Extended IP access list 106
10 permit icmp 172.16.199.0 0.0.0.127 any echo-reply
20 permit icmp 172.16.199.0 0.0.0.127 any ttl-exceeded
30 permit icmp 172.16.199.0 0.0.0.127 any unreachable
```

```
40 permit tcp 172.16.199.0 0.0.0.127 any established
50 deny ip 172.16.199.0 0.0.0.127 any
60 permit ip any any
Extended IP access list in-filter
10 permit icmp 172.16.199.0 0.0.0.127 any echo-reply
20 permit icmp 172.16.199.0 0.0.0.127 any ttl-exceeded
30 permit icmp 172.16.199.0 0.0.0.127 any unreachable
40 permit tcp 172.16.199.0 0.0.0.127 any established
50 deny ip 172.16.199.0 0.0.0.127 any
60 permit ip any any
```

注意：有 3 个 ACL，其中编号为 106 的 ACL 与命名为 in-filter 的 ACL 等价。

（20）在 Rt99c 上查看配置文件

```
Rt99c# show running-config
Building configuration...
Current configuration : 1512 bytes
……
!
access-list 2 deny 172.16.199.0 0.0.0.127
access-list 2 permit any
access-list 106 permit icmp 172.16.199.0 0.0.0.127 any echo-reply
access-list 106 permit icmp 172.16.199.0 0.0.0.127 any ttl-exceeded
access-list 106 permit icmp 172.16.199.0 0.0.0.127 any unreachable
access-list 106 permit tcp 172.16.199.0 0.0.0.127 any established
access-list 106 deny ip 172.16.199.0 0.0.0.127 any
access-list 106 permit ip any any
ip access-list extended in-filter
  permit icmp 172.16.199.0 0.0.0.127 any echo-reply
  permit icmp 172.16.199.0 0.0.0.127 any ttl-exceeded
  permit icmp 172.16.199.0 0.0.0.127 any unreachable
  permit tcp 172.16.199.0 0.0.0.127 any established
  deny ip 172.16.199.0 0.0.0.127 any
  permit ip any any
//以上 15 行显示已应用命名 ACL
!
……
```

注意：在应用命名为 in-filter 的 ACL 时自动取消应用编号为 106 的 ACL。

重新进行测试，效果应该与配置 ACL 106 时完全相同。

4．实训 4：测试验收

（1）在 PC99c 上浏览 http://www.168.com

如图 14-9 所示，在 PC99c 上浏览 http://www.168.com，测试 HTTP 服务。访问成功！

项目 14　ACL 基本配置

图 14-9　在 PC99c 上浏览 http://www.168.com

（2）在 PC99c 上浏览 http://www.pkt.net

如图 14-10 所示，在 PC99c 上浏览 http://www.pkt.net，测试 HTTP 服务。访问成功！

图 14-10　在 PC99c 上浏览 http://www.pkt.net

5．保存配置

通过全面测试，确认路由器的配置正确后，请保存配置，否则路由器重启后又要重新配置和调试。为了养成良好的习惯，每次完成模拟配置和调试后，要记着保存每台路由器的配置。在进行模拟实训时，最后还要记着保存网络拓扑。

单击【保存】按钮，保存配置到 P14002.pkt 文件中备用，然后再将其另存为文件名为 P14003.pkt 的文件。

学习总结

随着网络的扩大和开放，其面临的威胁越来越多，网络安全问题便成为网络管理员最头疼的问题。一方面，为了业务的发展，必须允许对网络资源的开放访问；另一方面，又必须确保数据和资源的尽可能安全。

确保网络安全可采用的技术很多，通过访问控制列表（ACL）对数据流进行过滤是实现网络安全的手段之一。

访问控制列表（ACL）使用包过滤技术，在路由器上读取第三层及第四层包头中的信息（如源地址、目的地址、源端口、目的端口等），根据预先定义好的规则对包进行过滤，从而达到访问控制的目的。

ACL 分很多种，不同场合应用不同种类的 ACL。

① 标准 ACL：最简单，用途有限，它只使用 IP 包中的源 IP 地址进行过滤，表号范围为 1～99 或 1300～1999。

② 扩展 ACL：比标准 ACL 具有更多的匹配项，功能更加强大和细化，可以针对包括协议类型、源地址、目的地址、源端口、目的端口、TCP 连接建立等进行过滤，表号范围为 100～199 或 2000～2699。

③ 命名 ACL：以列表名称代替列表编号来定义 ACL，同样包括标准和扩展两种访问控制列表。

此外，还有自反 ACL、动态 ACL 和基于时间的 ACL，但模拟器不支持。

请特别注意在访问控制列表中用到的以下两个术语。

① 通配符掩码或反掩码：一个 32 比特的二进制字符串，它规定了当一个 IP 地址与其他 IP 地址进行比较时，该 IP 地址中哪些位被忽略。通配符掩码中的"1"表示忽略 IP 地址中对应的位，而"0"则表示该位必须匹配。

两个极端的例子是：反掩码 255.255.255.255 表示忽略所有位，等价于关键字 any；反掩码 0.0.0.0 表示每一位都必须匹配，等价于关键字 host。

② Inbound 和 outbound：当在接口上应用访问控制列表时，用户要指明访问控制列表是应用于流入数据还是流出数据。

ACL 定义好以后，可以应用在很多地方，最常见的是应用在路由器接口上，其他的应用包括：在 NAT 中用于定义内部 IP 地址范围，在 vty 下用于控制 Telnet 访问（用 access-class 命令调用）等。

路由器每一个接口的每个方向，只能应用一个 ACL。入站 ACL 位于路由转发之前，即数据包先通过入站 ACL 过滤，再由路由进程转发或丢弃。出站 ACL 位于路由转发之后，即数据包先通过路由进程转发，再由出站 ACL 过滤。

访问控制列表表项的检查按自上而下的顺序进行，并且从第一个表项开始，一旦有匹配的表项，便不再检查后面的表项。所以配置访问控制列表的命令的次序很重要，必须慎重考虑，确保无误。

ACL 不对本路由器自身产生的 IP 数据包进行过滤，ACL 隐含的最后一条规则总是拒绝所有 IP 数据包。

标准 ACL 应尽量在靠近目的地的位置应用，因为标准 ACL 只使用源地址，如果将其靠近源会阻止数据包流向其他接口。而扩展 ACL 应尽量在靠近过滤源的位置应用，因为这样就不会影响其他接口上的数据流。

修改 ACL 很麻烦，因此，请先用文本编辑器编辑好命令，然后再复制、粘贴。

课后作业

完成上面的模拟实训，将实训过程的截图按顺序粘贴到一个 Word 文件里并用适当的

项目 14　ACL 基本配置

文字说明你对它的理解；总结本次实训所需要的主要命令及其作用，作为实训报告上交。

实训报告一律以"ID 姓名项目号.doc"为文件名命名，网络拓扑及其配置也以"ID 姓名项目号.pkt"为文件名保存并上交。例如，张三的 ID 为 03，他的文件名为"03 张三 14.doc"和"03 张三 14.pkt"。

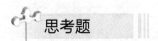

思考题

我们已经学习了路由器的很多应用，你对路由器的功能和用途，路由器的组成和工作原理都知道了吗？对 VLSM 和 CIDR 理解了吗？

请查阅有关资料，对上述问题做一个归纳总结。

至此，我们对交换机没做任何配置，交换机是否不需要配置呢？如果需要的话，它有哪些配置？如何配置？

在下一个项目中我们将学习相关知识。

项目 15 交换机及其基本配置

项目描述

若你是某单位的网络管理员，现在构建以太网一般都用交换机，为了确保网络中交换机的安全或者你想要通过远程登录方式来管理交换机，你需要对交换机做一些基本的配置。

网络拓扑

网络拓扑如图 15-1 所示。

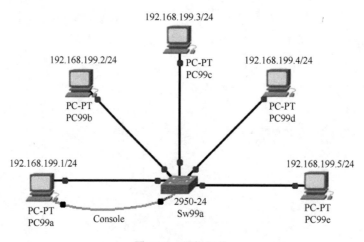

图 15-1 网络拓扑

学习目标

- 了解交换机的分类、外观和组成；
- 理解和掌握交换机的基本功能和原理；
- 理解和掌握以太网的构成和配置；
- 掌握交换机的基本配置；
- 会查看交换机的配置信息。

项目 15　交换机及其基本配置

实训任务分解

① 配置交换机。
② 用交换机组成星形局域网。
③ 测试配置效果。

知识点介绍

Cisco 公司的交换机从低端到高端有一系列的型号。大型交换网络一般多采用分层结构，可分为核心层、汇聚层和接入层三层。交换机各种型号适用于不同的组网需求，其性能和价格也有较大的差别。低端交换机用于接入层，高端交换机用于核心层，中端交换机一般用于汇聚层，也可用于核心层。

1．交换机的分类

交换机一般按其工作在 OSI 参考模型的对应层次，分为二层和三层交换机。如果没有特别说明，一般是指二层交换机。三层交换机集成了二层交换机和路由器的全部或部分功能。常见的 2924、2950 和 2960 是二层交换机，而 3550、3560 和 3750 是三层交换机。

2．交换机的组成部件

① CPU：中央处理器（Central Processing Unit），系统的核心部件，用于执行操作系统指令，实现高速数据传输。

② RAM/DRAM：随机存取存储器／动态随机存储器（Random Access Memory/Dynamic RAM），用于存储运行配置文件（running-config）。

③ NVRAM：非易失性（Non-volatile）RAM，用于存储启动配置文件（startup-config）。

④ Flash：闪存，非易失性存储器，可以以电子的方式存储和擦除，用于存储系统软件、启动配置等。

⑤ ROM：只读存储器（Read Only Memory），是一种永久性只能读取的存储器，用于存储开机诊断程序、引导程序。

⑥ 接口电路：各端口的内部电路。

⑦ IOS：Internetwork Operating System，管理交换机的操作系统软件。

3．交换机的作用和功能

交换机是网络集中设备，其端口连接各网络设备。交换机在以太网中起的作用是信息中转站。它把从某个端口接收到的数据从其他端口转发出去，在转发数据时，端口带宽是独享的。

交换机有以下三个基本功能：
- 地址学习；

- 转发或过滤；
- 避免环路。

4. 交换机的基本原理

为了进行数据帧的传输，连接在交换机端口上的主机通过地址解析协议（ARP）查询对方网卡的物理地址（MAC 地址）。交换机检查接收到的每个数据帧的源 MAC 地址，并将它与接收端口的对应关系记录在交换机的 MAC 地址表中，这就是所谓的地址学习。交换机在转发数据时，由二层数据帧结构中的目标 MAC 地址决定数据帧被转发到哪个端口。而路由器在转发数据时是根据目标 IP 地址来决定数据包转发到第二层哪个端口，并在二层封装成数据帧后发送出去。

交换机 MAC 地址表建立的过程也就是地址学习的过程，如图 15-2 所示。

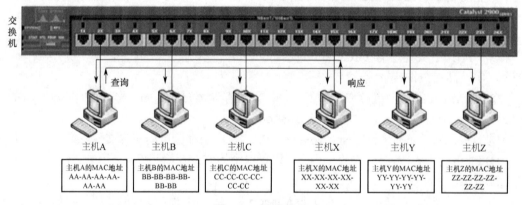

图 15-2　地址学习

（1）MAC 地址表的建立

① 主机 A 向目标主机 X 发送查询目标 MAC 地址信息，该信息会首先发送到交换机。

② 交换机在接收到查询数据帧后，会先将数据帧内的源 MAC 地址记录在自己的 MAC 地址表中，再向其他所有端口发送查询数据帧。

③ 若目标主机 X 接收到该信息后对源主机进行响应，则交换机就会将主机 X 的 MAC 地址也记录在自己的 MAC 地址表中。

④ 两台主机（主机 A 和主机 X）进行点对点的通信。

⑤ 如果两台主机在一定时间内未进行通信，交换机将会定时刷新自己的 MAC 地址表。

（2）MAC 地址表的路由过滤

当交换机接收到一个数据帧时，它会首先检查数据帧里的 MAC 地址。若目标地址尚未缓存在 MAC 地址表里，则交换机就向所有的其他端口发送数据帧（泛洪）；如果该地址已经缓存在 MAC 地址表里，交换机就会按照表中的地址进行转发，实现点对点通信，而不会广播或泛洪到其他端口。这称为交换机的 MAC 地址表的缓存过滤或路由过滤，它可减少对资源的占用，显著提高信息的交换速率。

交换机的数据转发模式如下：

- Cut-through（直通传送）；
- Store-and-forward（存储-转发）；
- Modified Cut-through（改进的直通传送）。

5. 实训环境和配置数据

实训环境如图 15-3 所示。

① 实训目的：
- 理解和掌握以太网的构成和配置；
- 掌握配置交换机的基本步骤和方法；
- 掌握交换机的基本配置；
- 会查看交换机的配置信息。

② 实训任务：
- 完成交换机的基本配置；
- 用交换机组成星形局域网；
- 测试验证配置效果。

③ 实训设备：
- 交换机 1 台（Catalyst 2900XL）；
- 带有网卡的 PC 5 台；
- 直连双绞线 5 条。

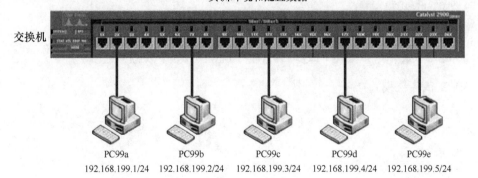

图 15-3 实训环境

与路由器配置相似，用反转线将交换机的 Console 端口与计算机的 COM 口相连接，在计算机上启动超级终端，登录交换机，进入特权模式，便可对交换机进行配置。

在模拟时，先构建图 15-1 所示的网络拓扑，再进行配置。

如果只是用交换机构成以太网，则只要接入交换机的设备的 IP 地址配置正确就行，无须对交换机做任何配置，即用其默认配置就行。

实训过程

1. 配置交换机

（1）完成交换机基本配置

```
Sw99a(config)#hostname Sw99a
Sw99a(config)#enable secret 99secret
Sw99a(config)#line con 0
Sw99a(config-line)#logging sync
Sw99a(config-line)#exec-timeout 0 0
Sw99a(config-line)#line vty 0 4    //配置 5 个 vty 会话，开启远程登录功能
Sw99a(config-line)#password 99vty0-4
Sw99a(config-line)#login
Sw99a(config-line)#logging sync
Sw99a(config-line)#exec-timeout 0 0
Sw99a(config-line)#exit
Sw99a(config)#service password-encr
Sw99a(config)#no ip domain-lookup
```

（2）配置远程登录用 IP 地址

注意：只有在需要远程登录时才配置 IP 地址。

```
Sw99a(config)#int vlan 1
Sw99a(config-if)#ip add 192.168.199.222 255.255.255.0
//配置远程登录用 IP 地址
Sw99a(config-if)#no sh
Sw99a(config-if)#exit
Sw99a(config)#ip default-gateway 192.168.199.254
//可以配置默认网关但没必要，即使与路由器相连也没必要配置默认网关
```

（3）测试特权模式口令

```
Sw99a>enable
Password:                          //输入密码 99secret
Sw99a#conf t
Sw99a(config)#
```

到此，说明设置的口令生效了。

2. 用交换机组成星形局域网

（1）配置 PC

在各台 PC 上配置 IP 地址、子网掩码和默认网关，然后单击【保存】按钮，保存已有配置。

① PC99a 的配置。
- IP 地址：192.168.199.1；
- 子网掩码：255.255.255.0；
- 默认网关：192.168.199.254。

② PC99b 的配置。
- IP 地址：192.168.199.2；
- 子网掩码：255.255.255.0；
- 默认网关：192.168.199.254。

③ PC99c 的配置。
- IP 地址：192.168.199.3；
- 子网掩码：255.255.255.0；
- 默认网关：192.168.199.254。

④ PC99d 的配置。
- IP 地址：192.168.199.4；
- 子网掩码：255.255.255.0；
- 默认网关：192.168.199.254。

⑤ PC99e 的配置。
- IP 地址：192.168.199.5；
- 子网掩码：255.255.255.0；
- 默认网关：192.168.199.254。

（2）从任何一台 PC 上远程登录 Sw99a

```
C:\>telnet 192.168.199.222
Trying 192.168.199.222 ...Open
User Access Verification
Password:                              //输入密码 99vty0-4
Sw99a>en
Password:                              //输入密码 99secret
Sw99a#conf t
Sw99a(config)#
```

这说明可用远程登录的方式管理交换机。

3. 测试配置效果

（1）在 Sw99a 上测试网络连通性

测试交换机到各主机之间的网络连通性。

```
Sw99a#ping 192.168.199.1
!!!!!
Sw99a#ping 192.168.199.2
.!!!!
Sw99a#ping 192.168.199.3
.!!!!
Sw99a#ping 192.168.199.4
.!!!!
Sw99a#ping 192.168.199.5
.!!!!
Sw99a#ping 192.168.199.254
//到网关不通，因为根本没接路由器
......
```

（2）在 PC99a 上测试网络连通性

测试各主机之间的网络连通性。

```
C:\>ping 192.168.199.2
!!!!!
C:\>ping 192.168.199.3
!!!!!
C:\>ping 192.168.199.4
!!!!!
C:\>ping 192.168.199.5
!!!!!
```

（3）保存配置

与路由器一样，通过全面测试，确认交换机的配置正确后，同样需要保存配置。

```
Sw99a#copy run start
```

然后，将相关配置保存到 P15001.pkt 文件中备用，再将其另存为文件名为 P15002.pkt 的文件。

测试说明：

主机 PC99a、PC99b、PC99c、PC99d 和 PC99e 之间都能互相 ping 通。

为了能用远程登录方式管理交换机，我们还给交换机配置了一个 IP 地址，因此从各主机都能远程登录交换机，当然也都能 ping 通交换机。

若对交换机没有进行任何配置，则实际使用的是交换机的默认配置，这样当 ping 192.168.199.222 时就不能 ping 通了。

思考：

如果主机 PC99a、PC99b、PC99c、PC99d 和 PC99e 的 IP 地址的网络号不同会怎样呢？譬如将 PC99b 的 IP 地址改为 192.168.88.2/24，将 PC99d 的 IP 地址改为 192.168.88.4/24，其他的配置都不变，结果会如何呢？

如图 15-4 和图 15-5 所示，分别修改 PC99b 和 PC99d 的 IP 地址。

项目 15 交换机及其基本配置

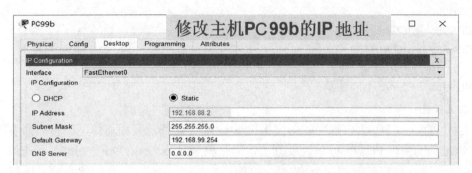

图 15-4 修改 PC99b 的 IP 地址

图 15-5 修改 PC99d 的 IP 地址

（4）在 Sw99a 上测试网络连通性

测试交换机到各主机之间的网络连通性。
Sw99a#ping 192.168.88.2
……
Sw99a#ping 192.168.88.4
……
到 PC99b 和 PC99d 不能通信。
Sw99a#ping 192.168.199.1
!!!!!
Sw99a#ping 192.168.199.3
!!!!!
Sw99a#ping 192.168.199.5
!!!!!
到 PC99a、PC99c、PC99e 仍然能通信。
C:\>ping 192.168.88.2   //在 PC99a 上测试
……
PC99a 与 PC99b 不能通信。
C:\>ping 192.168.199.3   //在 PC99a 上测试
!!!!!
PC99a 与 PC99c 仍然能通信。
C:\>ping 192.168.88.4   //在 PC99b 上测试

!!!!!
PC99b 与 PC99d 互相通信。
C:\>ping 192.168.199.5   //在 PC99b 上测试
......
PC99b 与 PC99e 不能通信。
测试说明：
主机 PC99a、PC99c、PC99e 与交换机 Sw99a 之间都能互相通信，但它们都不能与修改 IP 地址后的主机 PC99b 和 PC99d 互通，而主机 PC99b 与 PC99d 之间能互相通信。
由同一台交换机连接的两个 IP 网络之间不能通信，但两个 IP 网络内的主机都能互相通信，说明两个 IP 网络的主机被分成了两个网段或两个广播域。
查看交换机信息：
可以用 show 命令查看交换机的相关信息。
① Sw99a# show run   //显示配置信息
② Sw99a# show version   //显示版本信息
③ Sw99a# show flash   //显示 Flash
④ Sw99a# show vlan brief   //显示 VLAN 信息
⑤ Sw99a# show mac-   //显示 MAC 地址表
⑥ Sw99a# show processes   //显示进程信息
⑦ Sw99a# show span   //显示生成树信息

（5）显示配置信息

```
Sw99a#show run
Building configuration...
Current configuration : 1242 bytes
!
version 12.1
no service timestamps log datetime msec
no service timestamps debug datetime msec
service password-encryption
!
hostname Sw99a
!
enable secret 5 $1$mERr$3UwT5QPKcccm/zMsJ1niL1
!
no ip domain-lookup
!
spanning-tree mode pvst
spanning-tree extend system-id
!
interface FastEthernet0/1
!
interface FastEthernet0/2
……
```

项目 15 交换机及其基本配置

```
interface FastEthernet0/24
!
interface Vlan1
ip address 192.168.199.222 255.255.255.0
!
ip default-gateway 192.168.199.254
!
line con 0
logging synchronous
exec-timeout 0 0
!
line vty 0 4
exec-timeout 0 0
password 7 087815581D00555A46
logging synchronous
login
line vty 5 15
login
!
End
```

（6）显示版本信息

```
Sw99a#show version
Cisco Internetwork Operating System Software
IOS (tm) C2950 Software (C2950-I6Q4L2-M), Version 12.1(22)EA4, RELEASE SOFTWARE(fc1)
Copyright (c) 1986-2005 by Cisco Systems, Inc.
Compiled Wed 18-May-05 22:31 by jharirba
Image text-base: 0x80010000, data-base: 0x80562000


ROM: Bootstrap program is C2950 boot loader
……

63488K bytes of flash-simulated non-volatile configuration memory.
……
```

（7）显示 Flash 和 VLAN 信息

```
Sw99a#show flash
Directory of flash:/
1 -rw- 3058048 <no date> c2950-i6q4l2-mz.121-22.EA4.bin
2 -rw- 1242 <no date> config.text
64016384 bytes total (60957094 bytes free)


Sw99a#show vlan brief
VLAN  Name              Status    Ports
-----------------------------------------------------------------
1     default           active    Fa0/1, Fa0/2, Fa0/3, Fa0/4
```

Fa0/5, Fa0/6, Fa0/7, Fa0/8
Fa0/9, Fa0/10, Fa0/11, Fa0/12
Fa0/13, Fa0/14, Fa0/15, Fa0/16
Fa0/17, Fa0/18, Fa0/19, Fa0/20
Fa0/21, Fa0/22, Fa0/23, Fa0/24

（8）显示 MAC 地址表

Sw99a#show mac-address-table
Mac Address Table

Vlan	Mac Address	Type	Ports
1	0000.0c1e.0604	DYNAMIC	Fa0/1
1	0001.964d.79b2	DYNAMIC	Fa0/3
1	0001.96cb.5378	DYNAMIC	Fa0/4
1	000c.cf64.39a5	DYNAMIC	Fa0/2
1	0090.2132.bca7	DYNAMIC	Fa0/5

（9）显示进程信息

Sw99a#show processes
CPU utilization for five seconds: 0%/0%; one minute: 0%; five minutes: 0%
PID QTy PC Runtime (ms) Invoked uSecs Stacks TTY Process
1 Csp 602F3AF0 0 1627 0 2600/3000 0 Load Meter
2 Lwe 60C5BE00 4 136 29 5572/6000 0 CEF Scanner
……
43 Hwe 8040E338 0 2 0 2560/3000 0 STP FAST TRANSI

（10）显示生成树信息

Sw99a#show spanning-tree
VLAN0001
Spanning tree enabled protocol ieee
Root ID Priority 32769
Address 0060.3E51.9B3E
This bridge is the root
Hello Time 2 sec Max Age 20 sec Forward Delay 15 sec
Bridge ID Priority 32769 (priority 32768 sys-id-ext 1)
Address 0060.3E51.9B3E
Hello Time 2 sec Max Age 20 sec Forward Delay 15 sec
Aging Time 20
Interface Role Sts Cost Prio.Nbr Type
---------------- ---- --- --------- -------- --------------------------------
Fa0/1 Desg FWD 19 128.1 P2p
Fa0/2 Desg FWD 19 128.2 P2p
Fa0/3 Desg FWD 19 128.3 P2p
Fa0/4 Desg FWD 19 128.4 P2p
Fa0/5 Desg FWD 19 128.5 P2p

项目 15 交换机及其基本配置

学习总结

交换机与路由器的配置环境和基本配置都很相似。

在不需要用远程登录方式进行管理时，交换机和路由器都不用配置 vty，交换机也不用配置 IP 地址。同一个以太网中的所有主机的 IP 地址的网络号必须相同。

分属不同 IP 网络的主机即使连接在同一台交换机上也不能互相通信。要连通不同的 IP 网络，必须借助于路由功能。除了路由器，三层交换机也具有路由功能。

课后作业

完成上面的模拟实训，将实训过程的截图按顺序粘贴到一个 Word 文件里并用适当的文字说明你对它的理解；总结本次实训所需要的主要命令及其作用，作为实训报告上交。

实训报告一律以"ID 姓名项目号.doc"为文件名命名，网络拓扑及其配置也以"ID 姓名项目号.pkt"为文件名保存并上交。例如，张三的 ID 为 03，他的文件名为"03 张三 15.doc"和"03 张三 15.pkt"。

思考题

如果属于同一个部门的员工分布在不同的楼层，他们之间要互相通信、共享本部门资源，不同部门的员工之间的通信又要互相隔离，而且员工的分布可能会经常变动，那计算机网络如何构成和配置才便于管理呢？

在下一个项目中我们将学习有关的知识和技术。

项目 16 VLAN 及其配置

项目描述

某单位属于同一个部门的员工分布在不同的楼层，他们之间要互相通信、共享本部门资源，不同部门的员工之间的通信又要互相隔离，而且员工的办公位置可能会经常变动。若你是该单位的网络管理员，为了满足上述需求又便于你进行管理，你将如何构建该计算机网络？

网络拓扑

网络拓扑如图 16-1 所示。

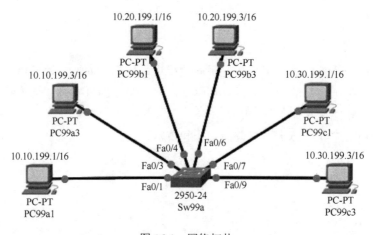

图 16-1 网络拓扑

学习目标

- 理解和掌握 VLAN 的概念和作用；
- 重点掌握 VLAN 的配置方法；
- 会查看交换机的相关信息；
- 会对网络进行测试验收。

项目 16　VLAN 及其配置

实训任务分解

① 创建 VLAN 10、VLAN 20 和 VLAN 30。
② 静态分配 VLAN 成员。
③ 检查和测试 VLAN 配置结果。

知识点介绍

1. VLAN 介绍

- VLAN：Virtual Local Area Network，虚拟局域网，是通过软件功能将物理交换机端口划分成一组逻辑上的设备或用户。
- 每个 VLAN 就像一个独立的物理网络；
- 同一个 VLAN 可以跨越多台交换机；
- 每个 VLAN 就像一个独立的 LAN；
- 主干功能支持多个 VLAN 间的数据传输。

VLAN 间通信需要三层设备（路由器），如果图 16-2 所示。

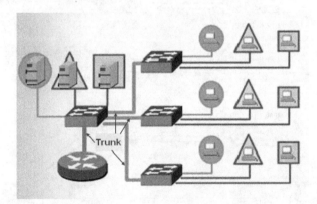

图 16-2　VLAN 间通信需要三层设备（路由器）

2. 为什么需要 VLAN

LAN 可以是由少数几台 PC 构成的小型网络，也可以是由数百台计算机构成的企业网络，但都是使用路由器分割的网络，也就是广播域。VLAN（Virtual LAN）翻译成中文是"虚拟局域网"，特指使用交换机分割的网络，也就是广播域。

广播域指的是广播帧（目标 MAC 地址全部为 1）所能传递到的范围，亦即能够直接通信的范围。其实，除了广播帧（Broadcast Frame），多播帧（Multicast Frame）和未知单播帧（Unknown Unicast Frame）也能在同一个广播域中畅行无阻。二层交换机本来只能构建单一的广播域，但使用 VLAN 技术以后，它就能够将网络分割成多个广播域。

如果仅有一个广播域，有可能会影响网络整体的传输性能。图 16-3 所示为广播帧传播示意图，在图 16-3 中，网络由 5 台二层交换机（交换机 1～5）及其连接的大量主机构成。假设主机 A 需要与主机 B 通信。在基于以太网的通信中，必须在数据帧中指定目标 MAC 地址才能正常通信，因此主机 A 必须先广播 ARP Request（ARP 请求）来尝试获取主机 B 的 MAC 地址。

交换机 1 收到广播帧（ARP 请求）后，会将它转发给除接收端口外的其他所有端口，也就是泛洪。接着交换机 2 收到广播帧后也会泛洪。交换机 3、4、5 也还会泛洪。最终 ARP 请求会被转发到同一网络中的所有客户机上。

请注意，这个 ARP 请求原本是为了获得计算机 B 的 MAC 地址而发出的，也就是说，只要计算机 B 能收到就万事大吉了。可事实上，数据帧却传遍了整个网络，导致所有的计算机都收到了它。如此一来，一方面广播信息消耗了网络整体的带宽；另一方面，收到广播信息的计算机还要消耗一部分 CPU 时间来对它进行处理，造成了网络带宽和 CPU 运算能力的大量无谓消耗。

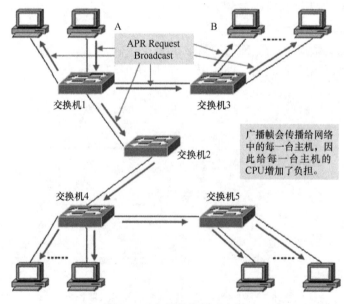

图 16-3　广播帧传播示意图

实际上，RIP、DHCP、NetBEUI、IPX、AppleTalk 等网络协议都采用广播通信，ARP 请求也采用广播通信。因此，在这些网络协议起作用的网络中都会频繁出现广播信息。

如果整个网络只有一个广播域，那么广播信息一旦发出，就会传遍整个网络，并给网络中的主机带来额外的负担。因此，在设计 LAN 时，需要注意如何才能有效地分割广播域。

在分割广播域时，一般都要使用路由器。使用路由器后，可以以路由器上的网络接口（LAN Interface）为单位分割广播域。使用路由器分割广播域的个数完全取决于路由器的网络接口个数。路由器通常不会有太多的网络接口，其数目为 1～4 个。因此，用户无法根据实际需要分割很多个广播域。

与路由器相比，二层交换机一般具有很多个网络接口。如果能使用它分割广播域，那么网络应用的灵活性会大大提高。用二层交换机分割广播域的技术，就是 VLAN。通过 VLAN，我们可以灵活地设计广播域，提高网络设计的自由度。

3. VLAN 的访问链接

什么是访问链接？

访问链接只属于一个 VLAN，且仅向该 VLAN 转发数据帧。

在大多数情况下，访问链接端口所连的是客户机。

访问链接既可以事先设定，也可以是根据所连的计算机的 MAC 地址或 IP 地址动态地设定。前者被称为"静态 VLAN"，后者即所谓的"动态 VLAN"。

交换机的端口可以分为以下两种：

- 访问链接端口；
- 汇聚链接端口。

设置 VLAN 的通常顺序：

- 生成 VLAN；
- 设定访问链接（决定各端口属于哪个 VLAN）。

4. VLAN 的汇聚（主干）链接

到此为止，我们学习的都是使用单台交换机设置 VLAN 的知识。那么，如果需要设置跨越多台交换机的 VLAN 又如何做呢？

图 16-4 给出了跨交换机 VLAN 直接连接示意图，在该图中，如果要将不同楼层的主机 A、C 和 B、D 各设置在同一个 VLAN 中，交换机 1 和交换机 2 该如何连接才好呢？最简单的方法自然是在交换机 1 和交换机 2 上各设一个 VLAN 专用的端口并互连。

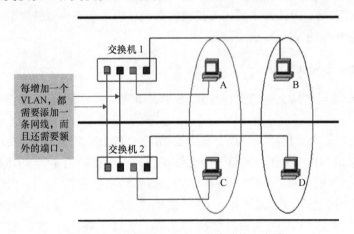

图 16-4　跨交换机 VLAN 直接连接示意图

但这个办法从扩展性和管理效率来看都不好。例如，当再新建 VLAN 时，就需要在交换机间连接新的网线。楼层间的布线是比较麻烦的，而且，VLAN 越多，交换机间互连所需要的端口也越多。交换机端口的利用率低是对资源的一种浪费，也限制了网络的扩展。

为了避免这种低效率的连接方式，人们想办法让交换机间互连的网线集中到一根上，这时使用的就是汇聚链接方式。

5. 汇聚链接

汇聚链接也称主干链接，指的是能够转发多个不同 VLAN 的数据帧的端口之间的链接。汇聚链路上传送的数据帧都被附加了用于识别分属于哪个 VLAN 的特殊信息。

现在再回过头来考虑一下刚才那个网络，如果采用汇聚链路又会如何呢？用户只需要简单地将交换机间互连的端口设定为汇聚链接端口就可以了。这时使用的还是普通的交叉网线，而不是什么其他的特殊线缆。

6. 汇聚方式

在交换机的汇聚链路上，通过对数据帧附加 VLAN 信息，构建跨越多台交换机的 VLAN。

最具有代表性的附加 VLAN 信息的技术有：
- IEEE 802.1q（dot1q）；
- ISL（Inter Switch Link）。

在通过汇聚链路时，附加 VLAN 识别信息有可能采用通用的 IEEE 802.1q 技术，也可能采用 Cisco 公司独有的 ISL 技术。

汇聚链路上既然要传送多个 VLAN 的数据，自然地负载就会较重。因此，设定的汇聚链接端口必须支持 100 Mbps 以上的传输速率。

默认条件下，汇聚链接会转发交换机上的所有 VLAN 的数据。换一个角度看，可以认为汇聚链接（端口）同时属于交换机上所有的 VLAN。由于实际应用中很可能并不需要转发所有 VLAN 的数据，因此为了减轻交换机的负载，也为了减少对带宽的浪费，我们可以对经由汇聚链路互连的 VLAN 设定限制。

7. ISL（Inter Switch Link）

ISL 是 Cisco 公司产品支持的一种与 IEEE 802.1q 类似的、用于在汇聚链路上附加 VLAN 信息的协议。

使用 ISL 后，每个数据帧头部都会被附加 26 字节的 ISL 包头（ISL Header），并且在帧尾带上通过对包括 ISL 包头在内的整个数据帧进行计算后得到的 4 字节 CRC 值。也就是说，总共增加了 30 字节的信息。如图 16-5 所示为数据帧结构示意图。

在使用 ISL 时，由于原数据帧及其 CRC 值都被完整地保留，因此当数据帧离开汇聚链路时，只要简单地去除 ISL 包头和新 CRC 值即可，无须重新计算 CRC 值。

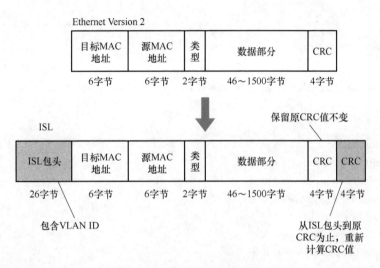

图 16-5　数据帧结构示意图

由于 ISL 是 Cisco 公司独有的协议，因此只能用于 Cisco 公司网络设备之间的互连。

8. IEEE 802.1q

IEEE 802.1q，英文缩写为 dot1q，俗称"Dot One Q"，是对数据帧附加 VLAN 识别信息的国际标准。

IEEE 802.1q 在以太网数据帧中"源 MAC 地址"与"类别域"（Type Field）之间附加 VLAN 识别信息，具体内容为 2 字节的 TPID（标签协议标识）和 2 字节的 TCI（标签控制信息），共计 4 字节。

数据帧在插入 TPID 和 TCI 后，CRC 值是对包括它们在内的整个数据帧重新计算后所得的值。而当数据帧离开汇聚链路时，TPID 和 TCI 会被去除，这时还会重新计算 CRC 值。

TPID 的值固定为 0x8100，交换机通过 TPID 来确定数据帧内附加了基于 IEEE 802.1q 的 VLAN 信息。而实质上的 VLAN ID 是 TCI 中的 12 位。由于总共有 12 位，因此最多可供识别 4096 个 VLAN。

基于 IEEE 802.1q 附加的 VLAN 信息，就像在邮递物品时附加的标签。因此，它也被称作"标签型 VLAN"（Tagging VLAN）。

汇聚链路（Trunk）只解决了交换机间多个 VLAN 的通信问题，不同 VLAN 之间仍是互不相通的。因此，在交换机上设置 VLAN 后，如果未进行其他处理，VLAN 间是无法通信的。

明明接在同一台交换机上，却又偏偏无法通信。这个事实也许让人难以接受，但它既是 VLAN 方便易用的特征，又是使 VLAN 令人费解的原因。当我们需要在不同的 VLAN 间进行通信时该怎么办？

VLAN 是广播域，而通常两个广播域之间由路由器连接，广播域之间来往的数据包都是由路由器中继的。因此，VLAN 间的通信也需要路由器提供中继服务，这被称作"VLAN 间路由"。VLAN 间路由，可以使用普通的路由器，也可以使用三层交换机。

实训过程

1. 创建 VLAN 10、VLAN 20、VLAN 30

（1）完成交换机基本配置

```
Sw99a(config)#hostname Sw99a
Sw99a(config)#enable secret 99secret
Sw99a(config)#service password-encr
Sw99a(config)#no ip domain-lookup
Sw99a(config)#line con 0
Sw99a(config-line)#logging sync
Sw99a(config-line)#exec-timeout 0 0
Sw99a(config-line)#line vty 0 4
Sw99a(config-line)#password 99vty0-4
Sw99a(config-line)#login
Sw99a(config-line)#logging sync
Sw99a(config-line)#exec-timeout 0 0
Sw99a(config-line)#end
```

（2）显示 VLAN 信息

```
Sw99a# show vlan brief
VLAN   Name                  Status    Ports
---------------------------------------------------------------
1      default               active    Fa0/1, Fa0/2, Fa0/3, Fa0/4
                                       Fa0/5, Fa0/6, Fa0/7, Fa0/8
                                       Fa0/9, Fa0/10, Fa0/11, Fa0/12
                                       Fa0/13, Fa0/14, Fa0/15, Fa0/16
                                       Fa0/17, Fa0/18, Fa0/19, Fa0/20
                                       Fa0/21, Fa0/22, Fa0/23, Fa0/24
1002   fddi-default          active
1003   token-ring-default    active
1004   fddinet-default       active
1005   trnet-default         active
```

（3）创建 VLAN

```
Sw99a(config)# vlan 10
Sw99a(config-vlan)# name VLANA
Sw99a(config)# vlan 20
Sw99a(config-vlan)# name VLANB
Sw99a(config)# vlan 30
Sw99a(config-vlan)# name VLANC
```

```
Sw99a(config-vlan)# end
```

（4）再次显示 VLAN 信息

```
Sw99a# show vlan brief
VLAN Name                Status     Ports
---------------------------------------------------------------
1    default             active     Fa0/1, Fa0/2, Fa0/3, Fa0/4
                                    Fa0/5, Fa0/6, Fa0/7, Fa0/8
                                    Fa0/9, Fa0/10, Fa0/11, Fa0/12
                                    Fa0/13, Fa0/14, Fa0/15, Fa0/16
                                    Fa0/17, Fa0/18, Fa0/19, Fa0/20
                                    Fa0/21, Fa0/22, Fa0/23, Fa0/24
10   VLANA               active
20   VLANB               active
30   VLANC               active
```

2. 静态分配 VLAN 成员

（1）把端口分配给 VLAN

```
Sw99a(config)#int ran fa0/1-3              //将端口 1~3 分配给 VLAN 10
Sw99a(config-if-range)#swi acc vlan 10
Sw99a(config-if-range)#int ran fa0/4-6     //将端口 4~6 分配给 VLAN 20
Sw99a(config-if-range)#swi acc vlan 20
Sw99a(config-if-range)#int ran fa0/7-9     //将端口 7~9 分配给 VLAN 30
Sw99a(config-if-range)#swi acc vlan 30
Sw99a(config-if-range)#end
```

（2）显示 VLAN 信息

```
Sw99a#show vlan brief
VLAN Name                Status     Ports
---------------------------------------------------------------
1    default             active     Fa0/10, Fa0/11, Fa0/12, Fa0/13
                                    Fa0/14, Fa0/15, Fa0/16, Fa0/17
                                    Fa0/18, Fa0/19, Fa0/20, Fa0/21
                                    Fa0/22, Fa0/23, Fa0/24
10   VLANA               active     Fa0/1, Fa0/2, Fa0/3
20   VLANB               active     Fa0/4, Fa0/5, Fa0/6
30   VLANC               active     Fa0/7, Fa0/8, Fa0/9
```

（3）配置 PC 的 IP 地址

在各台 PC 上配置 IP 地址、子网掩码和默认网关，然后单击【保存】按钮，保存已有配置。

① PC99a1。

- IP 地址：10.10.199.1；
- 子网掩码：255.255.0.0；
- 默认网关：10.10.0.1。

② PC99a3。
- IP 地址：10.10.199.3；
- 子网掩码：255.255.0.0；
- 默认网关：10.10.0.1。

③ PC99b1。
- IP 地址：10.20.199.1；
- 子网掩码：255.255.0.0；
- 默认网关：10.20.0.1。

④ PC99b3。
- IP 地址：10.20.199.3；
- 子网掩码：255.255.0.0；
- 默认网关：10.20.0.1。

⑤ PC99c1。
- IP 地址：10.30.199.1；
- 子网掩码：255.255.0.0；
- 默认网关：10.30.0.1。

⑥ PC99c3。
- IP 地址：10.30.199.3；
- 子网掩码：255.255.0.0；
- 默认网关：10.30.0.1。

3. 检查和测试 VLAN 配置结果

（1）在 PC99a1 上测试网络连通性

```
C:\>ping 10.10.199.3                //VLAN 内能通信
!!!!!
C:\>ping 10.20.199.1                //VLAN 间不通信
……
```

（2）在 PC99a3 上测试网络连通性

```
C:\>ping 10.10.199.1                //VLAN 内能通信
!!!!!
C:\>ping 10.30.199.3                //VLAN 间不通信
……
```

项目 16　VLAN 及其配置

（3）在 PC99b1 上测试网络连通性

```
C:\>ping 10.20.199.3              //VLAN 内能通信
!!!!!
C:\>ping 10.10.199.3              //VLAN 间不通信
……
```

（4）在 PC99b3 上测试网络连通性

```
C:\>ping 10.20.199.1              //VLAN 内能通信
!!!!!
C:\>ping 10.30.199.1              //VLAN 间不通信
……
```

（5）在 PC99c1 上测试网络连通性

```
C:\>ping 10.30.199.3              //VLAN 内能通信
!!!!!
C:\>ping 10.10.199.1              //VLAN 间不通信
……
```

（6）在 PC99c3 上测试网络连通性

```
C:\>ping 10.30.199.1              //VLAN 内能通信
!!!!!
C:\>ping 10.20.199.1              //VLAN 间不通信
……
```

单击【保存】按钮，保存配置到 P16001.pkt 文件中。

学习总结

Cisco 公司的交换机不仅具有二层交换功能，它还具有 VLAN 等功能。VLAN 技术可以使我们很容易地控制广播域的大小、有效地解决"广播风暴"问题，简化对网络的管理，提高网络的灵活性和安全性。

在出厂时交换机的所有端口都属于 VLAN 1。当给新建的 VLAN 分配端口时，相应的端口就从 VLAN 1 移到了新建的 VLAN 中。在交换机上划分 VLAN 之后，属于不同 VLAN 的主机之间就不能通信了。

当需要在不同的 VLAN 间通信时就需要借助于路由功能。

课后作业

完成上面的模拟实训，将实训过程的截图按顺序粘贴到一个 Word 文件里并用适当的文字说明你对它的理解；总结本次实训所需要的主要命令及其作用，作为实训报告上交。

实训报告一律以"ID 姓名项目号.doc"为文件名命名，网络拓扑及其配置也以"ID 姓名项目号.pkt"为文件名保存并上交。例如，张三的 ID 为 03，他的文件名为"03 张三 16.doc"和"03 张三 16.pkt"。

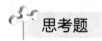思考题

当需要 VLAN 间通信时有哪些解决办法？
在下一个项目中我们将学习解决办法。

项目 17 VLAN 间路由与三层交换机配置

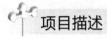项目描述

若你是某单位的网络管理员，你为各部门划分了不同的 VLAN，现在又需要各部门之间能互相通信，如果还想保留原来的 VLAN，那就需要借助于路由功能，或者将二层交换机与路由器的组合换成三层交换机。

在交换机上划分 VLAN 之后，属于不同 VLAN 的主机之间就不能通信了。

明明接在同一台交换机上，却又偏偏无法通信。这个事实也许让人难以接受，但它既是 VLAN 方便易用的特征，又是 VLAN 令人费解的原因。

那么，当我们需要不同的 VLAN 间相互通信时又该如何做呢？这就需要借助路由功能。实现 VLAN 间的路由，既可以使用普通的路由器，也可以使用三层交换机。

网络拓扑如图 17-1 所示。

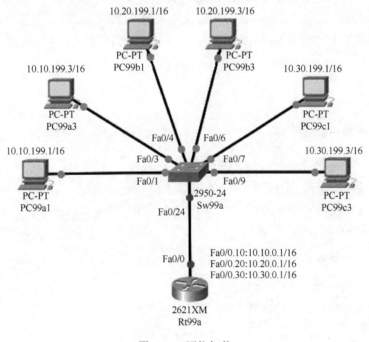

图 17-1 网络拓扑

学习目标

- 理解主干链路的概念和作用；
- 掌握主干链路的配置；
- 能用路由器连通多个 VLAN；
- 重点掌握三层交换机的 VLAN 配置；
- 理解三层交换机的功能和作用；
- 会对网络进行测试验收。

实训任务分解

① VLAN 间路由配置。
② 三层交换机的 VLAN 配置。

知识点介绍

如果项目 16 配置的 VLAN 之间需要互相通信，需要怎么做？

VLAN 之间通信需要借助于路由功能，因此，在交换机上需要接一台路由器。

打开文件 P16001.pkt，然后将其另存为文件名为 P17001.pkt 的文件备用，在项目 16 的网络拓扑中再添加一台路由器，如图 17-1 所示，这就是我们配置 VLAN 间路由的实训环境。

实训过程

1. VLAN 间路由配置

（1）在 Sw99a 上配置交换机的 Trunk 端口

```
Sw99a(config)#int fa0/24
Sw99a(config-if)#switch mode trunk        //把端口 24 配置为 trunk 模式
Sw99a(config-if)#end
```

（2）完成路由器的基本配置

```
Router(config)#host Rt99a
Rt99a(config)#enable secret 99secret
Rt99a(config)#service password-encr
Rt99a(config)#no ip domain-lookup
Rt99a(config)#line con 0
```

项目 17　VLAN 间路由与三层交换机配置

Rt99a(config-line)#logging sync
Rt99a(config-line)#exec-timeout 0 0
Rt99a(config-line)#line vty 0 4
Rt99a(config-line)#password 99vty0-4
Rt99a(config-line)#login
Rt99a(config-line)#logging sync
Rt99a(config-line)#exec-timeout 0 0
Rt99a(config-line)#exit

（3）配置路由器子接口

注意：IEEE 802.1q（dot1q）是虚拟局域网标准。

Rt99a(config-subif)#int fa0/0.10
Rt99a(config-subif)#encapsulation dot1q 10　　　　　　　　//VLAN 10 的封装模式为 dot1q
Rt99a(config-subif)#ip add 10.10.0.1 255.255.0.0
Rt99a(config-subif)#int fa0/0.20
Rt99a(config-subif)#encapsulation dot1q 20　　　　　　　　//VLAN 20 的封装模式为 dot1q
Rt99a(config-subif)#ip add 10.20.0.1 255.255.0.0
Rt99a(config-subif)#int fa0/0.30
Rt99a(config-subif)#encapsulation dot1q 30　　　　　　　　//VLAN 30 的封装模式为 dot1q
Rt99a(config-subif)#ip add 10.30.0.1 255.255.0.0

注意：各 VLAN 中主机的网关地址就是这里配置的 IP 地址。

（4）配置路由器接口

Rt99a(config-subif)#int fa0/0
Rt99a(config-if)#no ip add　　　　　　　　//取消 Fa0/0 的 IP 地址并激活 Fa0/0
Rt99a(config-if)#no sh

（5）显示 Trunk 端口信息

Sw99a#show interfaces trunk
Port Mode Encapsulation Status Native vlan
Fa0/24 on 802.1q trunking 1
Port Vlans allowed on trunk
Fa0/24 1-1005
Port Vlans allowed and active in management domain
Fa0/24 1,10,20,30
Port Vlans in spanning tree forwarding state and not pruned
Fa0/24 1,10,20,30

（6）显示路由信息

Rt99a# show ip route
10.0.0.0/16 is subnetted, 3 subnets

C 10.10.0.0 is directly connected, FastEthernet0/0.10
C 10.20.0.0 is directly connected, FastEthernet0/0.20
C 10.30.0.0 is directly connected, FastEthernet0/0.30

(7) 在主机上配置 IP 地址

在各台 PC 上配置 IP 地址、子网掩码和默认网关，然后单击【保存】按钮，保存已有配置。

① PC99a1 的配置。
- IP 地址：10.10.199.1；
- 子网掩码：255.255.0.0；
- 默认网关：10.10.0.1。

② PC99a3 的配置。
- IP 地址：10.10.199.3；
- 子网掩码：255.255.0.0；
- 默认网关：10.10.0.1。

③ PC99b1 的配置。
- IP 地址：10.20.199.1；
- 子网掩码：255.255.0.0；
- 默认网关：10.20.0.1。

④ PC99b3 的配置。
- IP 地址：10.20.199.3；
- 子网掩码：255.255.0.0；
- 默认网关：10.20.0.1。

⑤ PC99c1 的配置。
- IP 地址：10.30.199.1；
- 子网掩码：255.255.0.0；
- 默认网关：10.30.0.1。

⑥ PC99c3 的配置。
- IP 地址：10.30.199.3；
- 子网掩码：255.255.0.0；
- 默认网关：10.30.0.1。

(8) 在 Rt99a 上测试到各 VLAN 的网络连通性

```
Rt99a#ping 10.10.199.1                    //到 VLAN 10 通
!!!!!
Rt99a#ping 10.20.199.1                    //到 VLAN 20 通
!!!!!
```

```
Rt99a#ping 10.30.199.1                    //到 VLAN 30 通
!!!!!
```

（9）在 PC99a1 上测试 VLAN 间的网络连通性

```
C:\>ping 10.20.199.1
!!!!!
C:\>ping 10.30.199.1
!!!!!
```

（10）在 PC99b3 上测试 VLAN 间的网络连通性

```
C:\>ping 10.10.199.3
!!!!!
C:\>ping 10.30.199.3
!!!!!
```

（11）在 PC99c1 上测试 VLAN 间的网络连通性

```
C:\>ping 10.10.199.3
!!!!!
C:\>ping 10.20.199.3
!!!!!
```

测试表明，借助于路由器的路由功能，各 VLAN 之间都能通信了。验证配置正确后，请保存配置。在完成实训模拟时，除了保存配置，还要保存网络拓扑，然后将相关配置另存为文件名为 P17002.pkt 的文件备用。

2. 三层交换机的 VLAN 配置

在图 17-1 中，若 VLAN 之间的通信量很大，则 Sw99a 和 Rt99a 之间的主干链路就会成为通信"瓶颈"。如果将 Sw99a 和 Rt99a 集成在一个设备里，使 VLAN 之间的数据交换在硬件中实现，或达到"数据总线级"的速率，那就不存在通信瓶颈问题了。三层交换机就这样应运而生了。

我们可以把三层交换机看成一个集成了二层交换机功能和路由功能的设备，同样的需求可用三层交换机来实现。打开文件 P17002.pkt，在保存的网络拓扑中删除路由器，并把二层交换机 2950-24 换成三层交换机 3560-24PS，则得到的网络拓扑如图 17-2 所示。

（1）完成三层交换机基本配置

```
Switch(config)#hostname MS99a
MS99a(config)#enable secret 99secret
MS99a(config)#service password-encr
MS99a(config)#no ip domain-lookup
MS99a(config)#line con 0
MS99a(config-line)#logging sync
MS99a(config-line)#exec-timeout 0 0
```

MS99a(config-line)#line vty 0 4
MS99a(config-line)#password 99vty0-4
MS99a(config-line)#login
MS99a(config-line)#logging sync
MS99a(config-line)#exec-timeout 0 0
MS99a(config-line)#end

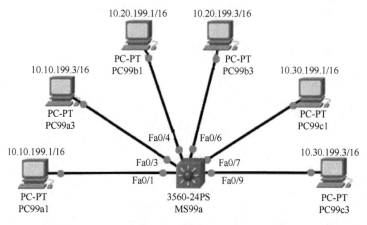

图 17-2　网络拓扑

（2）显示 VLAN 信息

```
MS99a#show vlan brief
VLAN Name                             Status     Ports
-------------------------------------------------------------------------------
1    default                          active     Fa0/1, Fa0/2, Fa0/3, Fa0/4
                                                 Fa0/5, Fa0/6, Fa0/7, Fa0/8
                                                 Fa0/9, Fa0/10, Fa0/11, Fa0/12
                                                 Fa0/13, Fa0/14, Fa0/15, Fa0/16
                                                 Fa0/17, Fa0/18, Fa0/19, Fa0/20
                                                 Fa0/21, Fa0/22, Fa0/23, Fa0/24
                                                 Gig0/1, Gig0/2
```

注意：初始时所有端口都属于 VLAN 1。

（3）创建 VLAN

```
MS99a(config)#vlan 10
MS99a(config-vlan)#name VLANA              //创建 VLAN 10，并命名为 VLANA
MS99a(config-vlan)#vlan 20
MS99a(config-vlan)#name VLANB              //创建 VLAN 20，并命名为 VLANB
MS99a(config-vlan)#vlan 30
MS99a(config-vlan)#name VLANC              //创建 VLAN 30，并命名为 VLANC
MS99a(config-vlan)#end
```

项目 17　VLAN 间路由与三层交换机配置

（4）显示 VLAN

```
MS99a#show vlan brief
VLAN Name                          Status    Ports
-------------------------------------------------------------------------
1    default                       active    Fa0/1, Fa0/2, Fa0/3, Fa0/4
                                             Fa0/5, Fa0/6, Fa0/7, Fa0/8
                                             Fa0/9, Fa0/10, Fa0/11, Fa0/12
                                             Fa0/13, Fa0/14, Fa0/15, Fa0/16
                                             Fa0/17, Fa0/18, Fa0/19, Fa0/20
                                             Fa0/21, Fa0/22, Fa0/23, Fa0/24
                                             Gig0/1, Gig0/2
10   VLANA                         active
20   VLANB                         active
30   VLANC                         active
//新建的 3 个 VLAN，尚未分配端口
```

（5）把端口分配给 VLAN

```
MS99a(config)#int ran fa0/1-3
MS99a(config-if-range)#swi acc vlan 10        //把端口 1～3 分配给 VLAN 10
MS99a(config-if-range)#int ran fa0/4-6
MS99a(config-if-range)#swi acc vlan 20        //把端口 4～6 分配给 VLAN 20
MS99a(config-if-range)#int ran fa0/7-9
MS99a(config-if-range)#swi acc vlan 30        //把端口 7～9 分配给 VLAN 30
MS99a(config-if-range)#end
```

（6）显示 VLAN

```
MS99a#show vlan brief
VLAN Name                          Status    Ports
-------------------------------------------------------------------------
1    default                       active    Fa0/10, Fa0/11, Fa0/12, Fa0/13
                                             Fa0/14, Fa0/15, Fa0/16, Fa0/17
                                             Fa0/18, Fa0/19, Fa0/20, Fa0/21
                                             Fa0/22, Fa0/23, Fa0/24, Gig0/1
                                             Gig0/2
10   VLANA                         active    Fa0/1, Fa0/2, Fa0/3
20   VLANB                         active    Fa0/4, Fa0/5, Fa0/6
30   VLANC                         active    Fa0/7, Fa0/8, Fa0/9
//端口从 VLAN 1 中移到了相应的 VLAN 中
```

（7）在主机上配置 IP 地址

各主机的 IP 地址在上面的 1 中已配置好，其中，
- VLANA 的子网地址为 10.10.0.0，网关地址为 10.10.0.1；

- VLANB 的子网地址为 10.20.0.0，网关地址为 10.20.0.1；
- VLANC 的子网地址为 10.30.0.0，网关地址为 10.30.0.1。

（8）在 PC99a1 上测试网络连通性

```
C:\>ping 10.10.199.3            //VLAN 内能通
!!!!!
C:\>ping 10.20.199.1            //VLAN 间不通
…..
```

（9）在 PC99b3 上测试网络连通性

```
C:\>ping 10.20.199.1            //VLAN 内能通
!!!!!
C:\>ping 10.30.199.3            //VLAN 间不通
…..
```

（10）在 PC99c1 上测试网络连通性

```
C:\>ping 10.30.199.3            //VLAN 内能通
!!!!!
C:\>ping 10.10.199.3            //VLAN 间不通
…..
```

（11）完成三层交换机的路由配置

```
MS99a(config)#ip routing                            //开启路由功能
MS99a(config)#int vlan 10
MS99a(config-if)#ip add 10.10.0.1 255.255.0.0       //配置 VLAN 10 端口的 IP 地址
MS99a(config-if)#int vlan 20
MS99a(config-if)#ip add 10.20.0.1 255.255.0.0       //配置 VLAN 20 端口的 IP 地址
MS99a(config-if)#int vlan 30
MS99a(config-if)#ip add 10.30.0.1 255.255.0.0       //配置 VLAN 30 端口的 IP 地址
MS99a(config-if)#end
```

注意：在各 VLAN 中主机的网关地址就是这里配置的 IP 地址。

（12）显示三层交换机的路由信息

```
MS99a#show ip route
Gateway of last resort is not set
10.0.0.0/16 is subnetted, 3 subnets
C 10.10.0.0 is directly connected, Vlan10
C 10.20.0.0 is directly connected, Vlan20
C 10.30.0.0 is directly connected, Vlan30
//以上 3 行表示只有 3 条直连路由，请注意每行最后面的端口名
```

（13）在 PC99a3 上测试网络连通性

```
C:\>ping 10.20.199.1                            //通
```

项目 17　VLAN 间路由与三层交换机配置

```
!!!!!
C:\>ping 10.30.199.1                              //通
!!!!!
```

（14）在 PC99b3 上测试网络连通性

```
C:\>ping 10.10.199.1                              //通
!!!!!
C:\>ping 10.30.199.1                              //通
!!!!!
```

（15）在 PC99c3 上测试网络连通性

```
C:\>ping 10.10.199.1                              //通
!!!!!
C:\>ping 10.20.199.1                              //通
!!!!!
```

测试表明，借助于三层交换机的路由功能，各 VLAN 之间都能通信。在交换机 MS99a 上执行保存配置命令 copy run start，然后先单击【保存】按钮保存相关配置，再将其另存为文件名为 P17003.pkt 的文件。

学习总结

VLAN 间的通信必须借助于路由功能，这既可以使用普通路由器实现，也可以使用三层交换机实现，但使用三层交换机实现效率会更高。

三层交换机是集成了路由功能和二层交换机功能的设备，因此也被称为路由交换机。本次实训很好地说明了三层交换机的功能和作用。

课后作业

完成上面的模拟实训，将实训过程的截图按顺序粘贴到一个 Word 文件里并用适当的文字说明你对它的理解；总结本次实训所需要的主要命令及其作用，作为实训报告上交。

实训报告一律以"ID 姓名项目号.doc"为文件名命名，网络拓扑及其配置也以"ID 姓名项目号.pkt"为文件名保存并上交。例如，张三的 ID 为 03，他的文件名为"03 张三 17.doc"和"03 张三 17.pkt"。

思考题

当 VLAN 涉及多个交换机时，如何简化 VLAN 的管理？
在下一个项目中我们将学习有关的知识和技术。

项目 18 VTP 及其配置

项目描述

若你是某单位的网络管理员，你要管理由多台交换机、多个 VLAN 构成的复杂交换网络，你如何简化管理？如果你发现一条主干链路的带宽不够，怎么办？

网络拓扑

网络拓扑如图 18-1 所示。

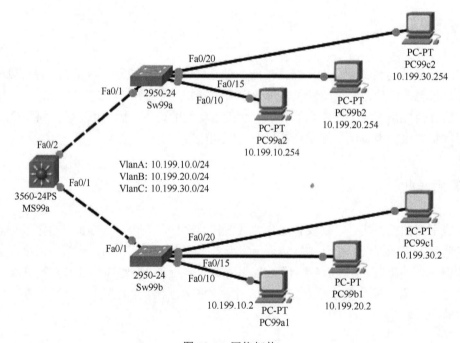

图 18-1 网络拓扑

学习目标

- 理解 VTP 的概念和作用；

项目 18　VTP 及其配置

- 重点掌握 VTP 的配置；
- 理解 EtherChannel 的概念和作用；
- 掌握 EtherChannel 的配置；
- 掌握三层交换机的配置；
- 会对网络进行测试验收。

实训任务分解

① VTP 配置。
② EtherChannel 配置。
③ VLAN 间通信配置。

知识点介绍

1. 关于 VTP

当 VLAN 涉及多台交换机时，如何简化 VLAN 的管理？

VTP（VLAN Trunk Protocol，VLAN 中继协议）就是用于简化对 VLAN 的管理的，它使我们可以只在配置为 VTP Server 模式的交换机上创建、修改、删除 VLAN，新的 VLAN 信息通过 Trunk 链路自动通告给其他交换机，任何采用 VTP 的交换机都可以接受这些修改，从而保持信息同步。但 VLAN 的端口分配不会自动传播，必须在每台交换机上进行配置，即交换机的端口分配是保持独立的。

VTP 分域来管理交换机，相同域中的交换机能共享 VLAN 信息。VTP 域中的交换机的工作模式有以下三种。

（1）服务器（Server）模式

在该模式下，在 VTP 服务器上能创建、修改、删除 VLAN，同时这些信息会被通告给域中的其他交换机。交换机的默认配置是服务器模式。每个 VTP 域必须至少有一台 VTP 服务器，当然也可以有多台 VTP 服务器。

（2）客户机（Client）模式

在该模式下，在 VTP 客户机上不允许创建、修改、删除 VLAN，但它会监听来自其他交换机的 VTP 通告并更改自己的 VLAN 信息，也会将接收到的 VTP 信息在 Trunk 链路上向其他交换机转发，因此这种交换机还能完成 VTP 中继。

（3）透明（Transparent）模式

在该模式下，交换机只会通过 Trunk 链路接收并转发 VTP 通告，充当 VTP 中继的角色，因此完全可以把该交换机看成透明的。在这种模式的交换机上也可以创建、修改、删除 VLAN，但是这些信息不会通告给其他交换机，它也不会用收到的其他交换机的 VTP 通告来更新自己的 VLAN 信息。

VTP 通告是以组播帧方式发送的，VTP 通告中有一个字段被称为修订号，初始值为 0。只要在 VTP Server 上创建、修改、删除 VLAN，通告中的修订号就增加 1，通告中还包含了 VLAN 的变化信息。修订号大的通告会覆盖修订号小的通告，而不管是 Server 还是 Client。交换机只接受比本地保存的修订号更大的通告；如果交换机收到比本地保存的修订号更小的通告，则会用自己的 VLAN 信息反向覆盖。

2. 关于 EtherChannel

EtherChannel（以太通道）是由 Cisco 公司开发的应用于交换机之间的多链路捆绑技术，它的基本原理如下所述。

将两个设备间多条快速或千兆位以太网物理链路捆绑在一起组成一条逻辑链路，从而达到带宽倍增的目的。除了增加带宽，EtherChannel 还可以在多条链路上均衡分配流量，起到负载分担的作用；当一条或多条链路发生故障时，只要还有链路正常工作，流量将转移到正常工作的链路上，整个过程在几毫秒内完成，从而起到冗余的作用，增强了网络的稳定性和安全性。

在 EtherChannel 中，负载在各个链路上的分布可以根据源 IP 地址、目的 IP 地址、源 MAC 地址、目的 MAC 地址、源 IP 地址和目的 IP 地址组合、源 MAC 地址和目的 MAC 地址组合等来完成。

两台交换机之间是否形成 EtherChannel 也可用协议自动协商。目前有两个协商协议：PAGP 和 LACP，PAGP 是 Cisco 公司专有的协议，LACP 是公共的标准协议。协商规律如表 18-1 所示。

表 18-1 协商规律

PAGP 协商规律总结				LACP 协商规律总结		
	ON	Desirable	auto		ON	active	passive
ON	√	×	×	ON	√	×	×
Desirable	×	√	√	active	×	√	√
auto	×	√	×	passive	×	√	×

3. VTP 配置

VTP 可以方便或简化对 VLAN 的管理，但在配置 VLAN 时，不一定要配置 VTP，它是可选项。请注意，交换机之间相连要用交叉线。

4. 实训环境和配置数据

实训环境如本项目的网络拓扑所示，MS99a 作为汇聚层交换机，Sw99a 和 Sw99b 作为接入层交换机（实际上接入层可能还有更多交换机）。现在要将主机分成三组，分别属于不同的 VLAN，配置数据表如表 18-2 所示。

项目 18　VTP 及其配置

表 18-2　配置数据表

VLAN			主　机	IP　地　址	子　网　掩　码
ID	Name	端　口			
10	VlanA	Fa0/10	PC99a1	10.199.10.2	255.255.255.0
		Fa0/10	PC99a2	10.199.10.254	255.255.255.0
20	VlanB	Fa0/15	PC99b1	10.199.20.2	255.255.255.0
		Fa0/15	PC99b2	10.199.20.254	255.255.255.0
30	VlanC	Fa0/20	PC99c1	10.199.30.2	255.255.255.0
		Fa0/20	PC99c2	10.99.30.254	255.255.255.0

实训过程

1. VTP 配置

（1）完成交换机 MS99a 的基本配置

```
Switch(config)#hostname MS99a
MS99a(config)#enable secret 99secret
MS99a(config)#service password-encr
MS99a(config)#no ip domain-lookup
MS99a(config)#line con 0
MS99a(config-line)#logging sync
MS99a(config-line)# exec-timeout 0 0
MS99a(config-line)#line vty 0 4
MS99a(config-line)#password 99vty0-4
MS99a(config-line)#login
MS99a(config-line)#logging sync
MS99a(config-line)#exec-timeout 0 0
MS99a(config-line)#end
```

（2）完成交换机 Sw99a 的基本配置

```
Switch(config)#hostname Sw99a
Sw99a(config)#enable secret 99secret
Sw99a(config)#service password-encr
Sw99a(config)#no ip domain-lookup
Sw99a(config)#line con 0
Sw99a(config-line)#logging sync
Sw99a(config-line)#exec-timeout 0 0
Sw99a(config-line)#line vty 0 4
Sw99a(config-line)#password 99vty0-4
Sw99a(config-line)#login
Sw99a(config-line)#logging sync
```

```
Sw99a(config-line)#exec-timeout 0 0
Sw99a(config-line)#end
```

(3) 完成交换机 Sw99b 的基本配置

```
Switch(config)#hostname Sw99b
Sw99b(config)#enable secret 99secret
Sw99b(config)#service password-encr
Sw99b(config)#no ip domain-lookup
Sw99b(config)#line con 0
Sw99b(config-line)#logging sync
Sw99b(config-line)#exec-timeout 0 0
Sw99b(config-line)#line vty 0 4
Sw99b(config-line)#password 99vty0-4
Sw99b(config-line)#login
Sw99b(config-line)#logging sync
Sw99b(config-line)#exec-timeout 0 0
Sw99b(config-line)#end
```

(4) 完成交换机 MS99a 的 VTP 配置

```
MS99a(config)#vtp version 2                              //启用版本 2
VTP mode already in V2.
MS99a(config)#vtp domain net08                           //配置 VTP 域
Domain name already set to net08.
MS99a(config)#vtp mode server                            //配置为 VTP 服务器
Device mode already VTP SERVER.
MS99a(config)#vtp password 99vtp                         //配置 VTP 口令
Setting device VLAN database password to 99vtp
MS99a(config)#end
```

(5) 完成交换机 Sw99b 的 VTP 配置

```
Sw99b(config)#vtp version 2                              //启用版本 2
Sw99b(config)#vtp domain net08                           //配置 VTP 域
Changing VTP domain name from NULL to net08
Sw99b(config)#vtp mode client                            //配置为 VTP 客户机
Setting device to VTP CLIENT mode.
Sw99b(config)#vtp password 99vtp                         //配置 VTP 口令
Setting device VLAN database password to 99vtp
Sw99b(config)#end
```

(6) 完成交换机 Sw99a 的 VTP 配置

```
Sw99a(config)#vtp version 2                              //启用版本 2
Sw99a(config)#vtp domain net08                           //配置 VTP 域
Changing VTP domain name from NULL to net08
```

项目 18　VTP 及其配置

```
Sw99a(config)#vtp mode client                              //配置为 VTP 客户机
Setting device to VTP CLIENT mode.
Sw99a(config)#vtp password 99vtp                           //配置 VTP 口令
Setting device VLAN database password to 99vtp
Sw99a(config)#end
```

（7）在 MS99a 上显示 VTP 状态信息

```
MS99a#show vtp status
VTP Version capable             : 1 to 2
VTP version running             : 2
VTP Domain Name                 : net08
VTP Pruning Mode                : Disabled
VTP Traps Generation            : Disabled
Device ID                       : 0090.0C0D.3400
Configuration last modified by 0.0.0.0 at 3-1-93 00:45:54
Local updater ID is 0.0.0.0 (no valid interface found)

Feature VLAN :
--------------
VTP Operating Mode              : Server
Maximum VLANs supported locally : 1005
Number of existing VLANs        : 5
Configuration Revision          : 0
MD5 digest                      : 0x0D 0x20 0xCB 0x27 0xA3 0x39 0xE8 0xB3
                                  0x7B 0xD8 0xB6 0xC1 0x92 0xEE 0xFD 0x31
```

（8）在 Sw99b 上显示 VTP 状态信息

```
Sw99b#show vtp status
VTP Version                     : 2
Configuration Revision          : 0
Maximum VLANs supported locally : 255
Number of existing VLANs        : 5
VTP Operating Mode              : Client
VTP Domain Name                 : net08
VTP Pruning Mode                : Disabled
VTP V2 Mode                     : Enabled
VTP Traps Generation            : Disabled
MD5 digest                      : 0x66 0xF7 0xD3 0x0D 0x18 0xAC 0x5B 0x57
```

（9）在 Sw99a 上显示 VTP 状态信息

```
Sw99a#show vtp status
VTP Version                     : 2
Configuration Revision          : 0
```

```
Maximum VLANs supported locally    : 255
Number of existing VLANs           : 5
VTP Operating Mode                 : Client
VTP Domain Name                    : net08
VTP Pruning Mode                   : Disabled
VTP V2 Mode                        : Enabled
VTP Traps Generation               : Disabled
MD5 digest                         : 0xCB 0xCB 0x11 0x4B 0x00 0x06 0x54 0x49
```

（10）在交换机 MS99a 上配置 Trunk 端口

```
MS99a(config)#int fa0/1
MS99a(config-if)#swi trunk encap dot1q          //配置 trunk 封装方式为 dot1q
MS99a(config-if)#swi mo tr
MS99a(config-if)#int fa0/2
MS99a(config-if)#swi trunk encap dot1q          //配置 trunk 封装方式为 dot1q
MS99a(config-if)#swi mo tr
```

注意：MS99a 的配置不能像 Sw99b 和 Sw99a 那样简单。

（11）在交换机 Sw99b 上配置 Trunk 端口

```
Sw99b(config)#int fa0/1
Sw99b(config-if)#swi mo tr
Sw99b(config-if)#end
```

注意：Sw99b 与 MS99a 的配置方法不同。

（12）在交换机 Sw99a 上配置 Trunk 端口

```
Sw99a(config)#int fa0/1
Sw99a(config-if)#swi mo tr
Sw99a(config-if)#end
```

注意：Sw99a 与 MS99a 的配置方法不同。

（13）在 MS99a 上显示 Trunk 端口信息

```
MS99a#show int trunk
Port      Mode         Encapsulation   Status         Native vlan
Fa0/1     on           802.1q          trunking       1
Fa0/2     on           802.1q          trunking       1

Port      Vlans allowed on trunk
Fa0/1     1-1005
Fa0/2     1-1005

Port      Vlans allowed and active in management domain
Fa0/1     1
Fa0/2     1

Port      Vlans in spanning tree forwarding state and not pruned
Fa0/1     1
```

| Fa0/2 | 1 |

(14) 在 Sw99b 上显示 Trunk 端口信息

```
Sw99b#show int trunk
Port        Mode          Encapsulation   Status      Native vlan
Fa0/1       on            802.1q          trunking    1
Port        Vlans allowed on trunk
Fa0/1       1-1005
Port        Vlans allowed and active in management domain
Fa0/1       1
Port        Vlans in spanning tree forwarding state and not pruned
Fa0/1       1
```

(15) 在 Sw99a 上显示 Trunk 端口信息

```
Sw99a#show int trunk
Port        Mode          Encapsulation   Status      Native vlan
Fa0/1       on            802.1q          trunking    1
Port        Vlans allowed on trunk
Fa0/1       1-1005
Port        Vlans allowed and active in management domain
Fa0/1       1
Port        Vlans in spanning tree forwarding state and not pruned
Fa0/1       1
```

(16) 在交换机 MS99a 上配置 VLAN

```
MS99a(config)#vlan 10
MS9a(config-vlan)#name VlanA           //创建 VLAN 10, 并命名为 VlanA
MS99a(config-vlan)#vlan 20
MS99a(config-vlan)#name VlanB          //创建 VLAN 20, 并命名为 VlanB
MS99a(config-vlan)#vlan 30
MS99a(config-vlan)#name VlanC          //创建 VLAN 30, 并命名为 VlanC
MS99a(config-vlan)#end
```

注意：只在 VTP 服务器 MS99a 上创建 VLAN。

(17) 在 MS99a 上显示 VLAN 信息

```
MS99a#show vlan brief
VLAN Name                           Status     Ports
---- -------------------------------- --------- -------------------------------
1    default                          active    Fa0/3, Fa0/4, Fa0/5, Fa0/6
                                                Fa0/7, Fa0/8, Fa0/9, Fa0/10
                                                Fa0/11, Fa0/12, Fa0/13, Fa0/14
                                                Fa0/15, Fa0/16, Fa0/17, Fa0/18
                                                Fa0/19, Fa0/20, Fa0/21, Fa0/22
```

			Fa0/23, Fa0/24, Gig0/1, Gig0/2
10	VlanA	active	
20	VlanB	active	
30	VlanC	active	

（18）在 Sw99b 上显示 VLAN 信息

```
Sw99b#show vlan brief
VLAN Name                          Status    Ports
---- -------------------------------- --------- -------------------------------
1    default                        active    Fa0/2, Fa0/3, Fa0/4, Fa0/5
                                              Fa0/6, Fa0/7, Fa0/8, Fa0/9
                                              Fa0/10, Fa0/11, Fa0/12, Fa0/13
                                              Fa0/14, Fa0/15, Fa0/16, Fa0/17
                                              Fa0/18, Fa0/19, Fa0/20, Fa0/21
                                              Fa0/22, Fa0/23, Fa0/24
10   VlanA                          active
20   VlanB                          active
30   VlanC                          active
```

（19）在 Sw99a 上显示 VLAN 信息

```
Sw99a#show vlan brief
VLAN Name                          Status    Ports
---- -------------------------------- --------- -------------------------------
1    default                        active    Fa0/2, Fa0/3, Fa0/4, Fa0/5
                                              Fa0/6, Fa0/7, Fa0/8, Fa0/9
                                              Fa0/10, Fa0/11, Fa0/12, Fa0/13
                                              Fa0/14, Fa0/15, Fa0/16, Fa0/17
                                              Fa0/18, Fa0/19, Fa0/20, Fa0/21
                                              Fa0/22, Fa0/23, Fa0/24
10   VlanA                          active
20   VlanB                          active
30   VlanC                          active
```

（20）在 Sw99b 上把端口分配给 VLAN

```
Sw99b(config)#int ran fa0/10-14                 //把端口 10～14 分配给 VLAN 10
Sw99b(config-if-range)#swi acc vlan 10
Sw99b(config-if-range)#int ran fa0/15-19        //把端口 15～19 分配给 VLAN 20
Sw99b(config-if-range)#swi acc vlan 20
Sw99b(config-if-range)#int ran fa0/20-24        //把端口 20～24 分配给 VLAN 30
Sw99b(config-if-range)#swi acc vlan 30
Sw99b(config-if-range)#end
```

（21）在 Sw99a 上把端口分配给 VLAN

```
Sw99a(config)#int ran fa0/10-14                 //把端口 10～14 分配给 VLAN 10
```

```
Sw99a(config-if-range)#swi acc vlan 10
Sw99a(config-if-range)#int ran fa0/15-19        //把端口 15~19 分配给 VLAN 20
Sw99a(config-if-range)#swi acc vlan 20
Sw99a(config-if-range)#int ran fa0/20-24        //把端口 20~24 分配给 VLAN 30
Sw99a(config-if-range)#swi acc vlan 30
Sw99a(config-if-range)#end
```

（22）在 MS99a 上显示 VLAN 信息

```
MS99a#show vlan brief
VLAN Name                         Status        Ports
---- -------------------------    ---------    -------------------------------
1    default                      active        Fa0/3, Fa0/4, Fa0/5, Fa0/6
                                                Fa0/7, Fa0/8, Fa0/9, Fa0/10
                                                Fa0/11, Fa0/12, Fa0/13, Fa0/14
                                                Fa0/15, Fa0/16, Fa0/17, Fa0/18
                                                Fa0/19, Fa0/20, Fa0/21, Fa0/22
                                                Fa0/23, Fa0/24, Gig0/1, Gig0/2
10   VlanA                        active
20   VlanB                        active
30   VlanC                        active
```

注意：交换机 MS99a 不用为新建的 VLAN 分配端口。

（23）在 Sw99b 上显示 VLAN 信息

```
Sw99b#show vlan brief
VLAN Name                         Status        Ports
---- -------------------------    ---------    -------------------------------
1    default                      active        Fa0/2, Fa0/3, Fa0/4, Fa0/5
                                                Fa0/6, Fa0/7, Fa0/8, Fa0/9
10   VlanA                        active        Fa0/10, Fa0/11, Fa0/12, Fa0/13, Fa0/14
20   VlanB                        active        Fa0/15, Fa0/16, Fa0/17, Fa0/18, Fa0/19
30   VlanC                        active        Fa0/20, Fa0/21, Fa0/22, Fa0/23, Fa0/24
```

（24）在 Sw99a 上显示 VLAN 信息

```
Sw99a#show vlan brief
VLAN Name                         Status        Ports
---- -------------------------    ---------    -------------------------------
1    default                      active        Fa0/2, Fa0/3, Fa0/4, Fa0/5
                                                Fa0/6, Fa0/7, Fa0/8, Fa0/9
10   VlanA                        active        Fa0/10, Fa0/11, Fa0/12, Fa0/13, Fa0/14
20   VlanB                        active        Fa0/15, Fa0/16, Fa0/17, Fa0/18, Fa0/19
30   VlanC                        active        Fa0/20, Fa0/21, Fa0/22, Fa0/23, Fa0/24
```

（25）完成各 PC 的配置

在各台 PC 上配置 IP 地址、子网掩码和默认网关，然后单击【保存】按钮，保存已有配置。

① PC99a1 的配置。
- IP 地址：10.199.10.2；
- 子网掩码：255.255.255.0；
- 默认网关：10.199.10.1。

② PC99a2 的配置。
- IP 地址：10.199.10.254；
- 子网掩码：255.255.255.0；
- 默认网关：10.199.10.1。

③ PC99b1 的配置。
- IP 地址：10.199.20.2；
- 子网掩码：255.255.255.0；
- 默认网关：10.199.20.1。

④ PC99ab2 的配置。
- IP 地址：10.199.20.254；
- 子网掩码：255.255.255.0；
- 默认网关：10.199.20.1。

⑤ PC99c1 的配置。
- IP 地址：10.199.30.2；
- 子网掩码：255.255.255.0；
- 默认网关：10.199.30.1。

⑥ PC99c2 的配置。
- IP 地址：10.199.30.254；
- 子网掩码：255.255.255.0；
- 默认网关：10.199.30.1。

（26）在 PC99a1 上测试连通性

```
C:\>ping 10.199.10.254                //VLAN 内能通
!!!!!
C:\>ping 10.199.20.254                //VLAN 间不通
…..
```

（27）在 PC99b2 上测试连通性

```
C:\>ping 10.199.20.2                  //VLAN 内能通
!!!!!
C:\>ping 10.199.30.2                  //VLAN 间不通
…..
```

（28）在 PC99c1 上测试连通性

```
C:\>ping 10.199.30.254                //VLAN 内能通
```

项目 18　VTP 及其配置

!!!!!
C:\>ping 10.199.10.2   //VLAN 间不通
……

在各交换机上执行保存配置命令 copy run start，保存配置到文件 P18001.pkt 中，再将其另存为文件名为 P18002.pkt 的文件，然后继续进行实训工作。

2. EtherChannel 配置

如果 Trunk 链路的带宽不够，怎么办？ EtherChannel 可以将多条物理链路捆绑在一起构成一条逻辑链路，从而使带宽倍增。

除了增加带宽，EtherChannel 还可在多条链路上均衡分配流量，起到负载分担的作用；在链路发生故障时，只要还有链路正常工作，还能起到冗余备份的作用。

EtherChannel 的配置环境如图 18-2 所示。

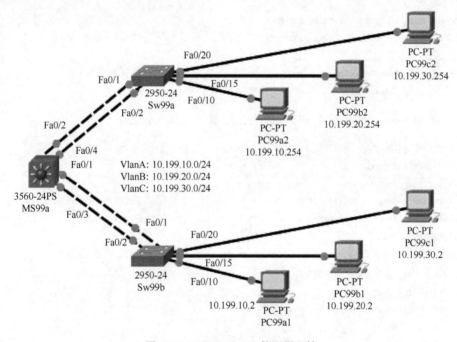

图 18-2　EtherChannel 的配置环境

（1）配置 EtherChannel 1 前，在 MS99a 上显示 EtherChannel 信息

```
MS99a#show etherchannel summary
Flags:  D - down         P - in port-channel
        I - stand-alone  s - suspended
        H - Hot-standby (LACP only)
        R - Layer3       S - Layer2
        U - in use       f - failed to allocate aggregator
        u - unsuitable for bundling
        w - waiting to be aggregated
```

```
            d - default port
Number of channel-groups in use: 0
Number of aggregators:         0
Group  Port-channel  Protocol   Ports
------+-------------+----------+-----------------------------------
```

（2）在 MS99a（Fa0/1 和 Fa0/3 端口）上配置 EtherChannel 1

```
MS99a(config)#int port-channel 1                //配置 EtherChannel 1
MS99a(config-if)#swi trunk encap dot1q
MS99a(config-if)#swi mo tr
MS99a(config-if)#int fa0/1
MS99a(config-if)#swi trunk encap dot1q
MS99a(config-if)#channel-group 1 mode on        //配置 EtherChannel 1 端口
MS99a(config-if)#int fa0/3
MS99a(config-if)#swi trunk encap dot1q
MS99a(config-if)#channel-group 1 mode on        //配置 EtherChannel 1 端口
MS99a(config-if)#exit
MS99a(config)#port-channel load-balance dst-mac //配置负载均衡方式
MS99a(config)#end
```

（3）在 Sw99b（Fa0/1 和 Fa0/2 端口）上配置 EtherChannel 1

```
Sw99b(config)#int port-channel 1                //配置 EtherChannel 1
Sw99b(config-if)#swi mo tr
Sw99b(config-if)#int fa0/1
Sw99b(config-if)#swi mo t                       //配置 Fa0/1 为主干模式
Sw99b(config-if)#channel-group 1 mo on          //配置 EtherChannel 1 端口
Sw99b(config-if)#int fa0/2
Sw99b(config-if)#swi mo tr                      //配置 Fa0/2 为主干模式
Sw99b(config-if)#channel-group 1 mo on          //配置 EtherChannel 1 端口
Sw99b(config-if)#ex
Sw99b(config)#port-channel load-balance dst-mac //配置负载均衡方式
Sw99b(config)#end
```

（4）配置 EtherChannel 1 后，在 MS99a 上显示 EtherChannel 信息

```
MS99a#show etherchannel summary
Flags:   D - down         P - in port-channel
         I - stand-alone  s - suspended
         H - Hot-standby (LACP only)
         R - Layer3       S - Layer2
         U - in use       f - failed to allocate aggregator
         u - unsuitable for bundling
         w - waiting to be aggregated
         d - default port
```

```
Number of channel-groups in use: 1
Number of aggregators:           1
Group   Port-channel   Protocol     Ports
------+--------------+-----------+---------------------------------------------
1       Po1(SU)         -          Fa0/1(P) Fa0/3(P)
```

（5）配置 EtherChannel 1 后，在 Sw99b 上显示 EtherChannel 信息

```
Sw99b#show eth sum
Flags:  D - down         P - in port-channel
        I - stand-alone  s - suspended
        H - Hot-standby (LACP only)
        R - Layer3       S - Layer2
        U - in use       f - failed to allocate aggregator
        u - unsuitable for bundling
        w - waiting to be aggregated
        d - default port
Number of channel-groups in use: 1
Number of aggregators:           1

Group   Port-channel   Protocol     Ports
------+--------------+-----------+---------------------------------------------
1       Po1(SU)         -          Fa0/1(P) Fa0/2(P)
```

（6）配置 EtherChannel 2 前，在 Sw99a 上显示 EtherChannel 信息

```
Sw99a#show eth sum
Flags: D - down P - in port-channel
I - stand-alone s - suspended
H - Hot-standby (LACP only)
R - Layer3 S - Layer2
U - in use f - failed to allocate aggregator
u - unsuitable for bundling
w - waiting to be aggregated
d - default port
Number of channel-groups in use: 0
Number of aggregators: 0
Group Port-channel Protocol Ports
------+--------------+-----------+------------------------------------
```

（7）在 MS99a（Fa0/2 和 Fa0/4 端口）上配置 EtherChannel 2

```
MS99a(config)#int port-channel 2                    //配置 EtherChannel 2
MS99a(config-if)#swi trunk encap dot1q
MS99a(config-if)#swi mo tr
MS99a(config-if)#int fa0/2
```

```
MS99a(config-if)#swi trunk encap dot1q
MS99a(config-if)#swi mo tr
MS99a(config-if)#channel-group 2 mode on          //配置 EtherChannel 2 端口
MS99a(config-if)#int fa0/4
MS99a(config-if)#swi trunk encap dot1q
MS99a(config-if)#swi mo tr
MS99a(config-if)#channel-group 2 mode on          //配置 EtherChannel 2 端口
MS99a(config-if)#exit
MS 99a(config)#port-channel load-balance dst-mac  //配置负载均衡方式
MS99a(config)#end
```

（8）在 Sw99a（Fa0/1 和 Fa0/2 端口）上配置 EtherChannel 2

```
Sw99a(config)#int port-channel 2                  //配置 EtherChannel 2
Sw99a(config-if)#swi mo tr
Sw99a(config-if)#int fa0/1
Sw99a(config-if)#swi mo tr
Sw99a(config-if)#channel-group 2 mo on            //配置 EtherChannel 2 端口
Sw99a(config-if)#int fa0/2
Sw99a(config-if)#swi mo tr
Sw99a(config-if)#channel-group 2 mo on            //配置 EtherChannel 2 端口
Sw99a(config-if)#exit
Sw99a(config)#port-channel load-balance dst-mac   //配置负载均衡方式
Sw99a(config)#end
```

（9）配置 EtherChannel 2 后，在 MS99a 上显示 EtherChannel 信息

```
MS99a#show etherchannel summary
Flags:  D - down         P - in port-channel
        I - stand-alone  s - suspended
        H - Hot-standby (LACP only)
        R - Layer3       S - Layer2
        U - in use       f - failed to allocate aggregator
        u - unsuitable for bundling
        w - waiting to be aggregated
        d - default port
Number of channel-groups in use: 2
Number of aggregators:           2

Group  Port-channel  Protocol    Ports
------+-------------+-----------+-----------------------------------------------
1      Po1(SU)         -         Fa0/1(P) Fa0/3(P)
2      Po2(SU)         -         Fa0/2(P) Fa0/4(P)
```

（10）配置 EtherChannel 2 后，在 Sw99b 上显示 EtherChannel 信息

```
Sw99b#show eth sum
```

```
Flags:  D - down         P - in port-channel
        I - stand-alone s - suspended
        H - Hot-standby (LACP only)
        R - Layer3       S - Layer2
        U - in use       f - failed to allocate aggregator
        u - unsuitable for bundling
        w - waiting to be aggregated
        d - default port
Number of channel-groups in use: 1
Number of aggregators:           1
Group  Port-channel  Protocol    Ports
------+-------------+-----------+-----------------------------------------------
1      Po1(SU)          -        Fa0/1(P) Fa0/2(P)
```

（11）配置 EtherChannel 2 后，在 Sw99a 上显示 EtherChannel 信息

```
Sw99a#show eth sum
Flags:  D - down         P - in port-channel
        I - stand-alone s - suspended
        H - Hot-standby (LACP only)
        R - Layer3       S - Layer2
        U - in use       f - failed to allocate aggregator
        u - unsuitable for bundling
        w - waiting to be aggregated
        d - default port
Number of channel-groups in use: 1
Number of aggregators:           1
Group  Port-channel  Protocol    Ports
------+-------------+-----------+-----------------------------------------------
2      Po2(SU)          -        Fa0/1(P) Fa0/2(P)
```

（12）在 PC99a2 上测试连通性

```
C:\>ping 10.199.10.2                          //VLAN 内能通
!!!!!
C:\>ping 10.199.20.2                          //VLAN 间不通
…..
```

（13）在 PC99b2 上测试连通性

```
C:\>ping 10.199.20.2                          //VLAN 内能通
!!!!!
C:\>ping 10.199.30.2                          //VLAN 间不通
…..
```

（14）在 PC99c2 上测试连通性

```
C:\>ping 10.199.30.2                              //VLAN 内能通
!!!!!
C:\>ping 10.199.10.2                              //VLAN 间不通
…..
```

3. VLAN 间通信配置

要想使 VLAN 之间能互相通信，在本例中只要在三层交换机 MS99a 的 VLAN 端口上配置 IP 地址即可。

（1）在 MS99a 上查看路由信息

```
MS99a#show ip route
Gateway of last resort is not set
```

（2）在 MS99a 的 VLAN 端口上配置 IP 地址

```
MS99a(config)#int vlan 10
MS99a(config-if)#ip add 10.199.10.1 255.255.255.0     //VLAN 10 网关地址
MS99a(config-if)#int vlan 20
MS99a(config-if)#ip add 10.199.20.1 255.255.255.0     //VLAN 20 网关地址
MS99a(config-if)#int vlan 30
MS99a(config-if)#ip add 10.199.30.1 255.255.255.0     //VLAN 30 网关地址
MS99a(config-if)#end
```

（3）再次在 MS99a 上查看路由信息

```
MS99a#show ip route
Gateway of last resort is not set
10.0.0.0/24 is subnetted, 3 subnets
C 10.199.10.0 is directly connected, Vlan10
C 10.199.20.0 is directly connected, Vlan20
C 10.199.30.0 is directly connected, Vlan30
```

（4）在 PC99a1 上测试连通性

```
C:\>ping 10.199.20.2                              //VLAN 间能通
!!!!!
C:\>ping 10.199.30.2                              //VLAN 间能通
!!!!!
```

（5）保存配置

测试表明，VTP、VLAN 和 EtherChannel 的配置都正确。

在各交换机上执行保存配置命令 copy run start，保存配置到文件 P18002.pkt 中，然后

项目 18　VTP 及其配置

再将其另存为文件名为 P18003pkt 的文件。

学习总结

VTP 可以方便或简化对 VLAN 的管理，它使我们可以只在配置为 VTP Server 模式的交换机上创建、修改、删除 VLAN，新的 VLAN 信息会通过 Trunk 链路自动通告给其他交换机，任何采用 VTP 的交换机都可以接受这些修改，从而保持信息同步。但 VLAN 的端口分配不会自动传播，必须在每台交换机上进行配置，各交换机对端口的分配仍保持独立性。

EtherChannel 可以将多条物理链路捆绑在一起构成一条逻辑链路，从而使带宽倍增。

交换机之间相连要用交叉线。

课后作业

完成上面的模拟实训，将实训过程的截图按顺序粘贴到一个 Word 文件里并用适当的文字说明你对它的理解；总结本次实训所需要的主要命令及其作用，作为实训报告上交。

实训报告一律以 "ID 姓名项目号.doc" 为文件名命名，网络拓扑及其配置也以 "ID 姓名项目号.pkt" 为文件名保存并上交。例如，张三的 ID 为 03，他的文件名为 "03 张三 18.doc" 和 "03 张三 18.pkt"。

思考题

冗余链路常用于提高网络的可靠性，但会使物理网络形成环路而产生广播风暴、多帧副本和 MAC 地址表不稳定的问题，怎么解决这些问题呢？

在下一个项目中我们将学习有关的知识和技术。

项目 19 STP 及其配置

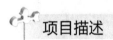

若你是某单位的网络管理员，为了提高网络的可靠性，你在交换网络中要增加一些冗余链路。但冗余链路会使物理网络形成环路而产生广播风暴、多帧复制和交换机 CAM 表不稳定的问题，你怎么解决这些问题？

生成树协议（STP）用于消除网络中的环路，使一个有环路的物理交换网络变成一个无环路的树形逻辑网络。

网络拓扑

网络拓扑如图 19-1 所示。

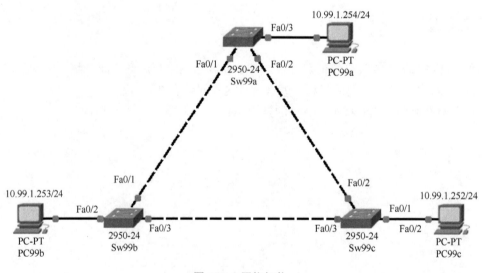

图 19-1 网络拓扑

学习目标

- 理解 STP 的概念和作用；

项目 19　STP 及其配置

- 重点掌握 STP 的配置；
- 会查看 STP 的状态；
- 会对网络进行测试验收。

实训任务分解

① 交换机基本配置。
② PC 配置。
③ 网络连通性测试。
④ 生成树信息显示。

知识点介绍

1．冗余链路

为了消除单点故障，提高网络的健壮性（可靠性）和稳定性，实际的网络中通常都使用一些备份连接。备份连接（也称备份链路）就是冗余链路。

当主链路出故障时，冗余链路自动启用，从而可提高网络的整体可靠性。

使用冗余链路能够为网络带来许多好处，但是冗余链路会使网络产生环路。环路问题是冗余链路所产生的最为严重的问题，它直接产生以下麻烦：

- 广播风暴；
- 多帧复制；
- 交换机 CAM 表不稳定。

2．生成树协议（STP）

STP 的全称是 Spanning Tree Protocol。

STP 协议是一个二层的链路管理协议，它在提供冗余链路的同时防止网络产生环路。

STP 有几种不同的版本，而且是互不兼容的。因此所有的网桥和交换机使用相同的 STP 版本非常重要。

3．生成树协议的发展

生成树协议和其他协议一样，随着网络的不断发展而不断更新换代。在生成树协议发展过程中，缺陷不断被克服，新特性不断被开发出来。按照功能的改进情况，我们可以粗略地把生成树协议划分为以下三代：

- 第一代生成树协议：STP/RSTP；
- 第二代生成树协议：PVST/PVST；
- 第三代生成树协议：MISTP/MSTP。

4. STP 的目的和作用

生成树协议的目的和作用是在保证提供冗余链路的前提下避免产生环路，它将有环路的物理网络变为无环路的树形逻辑网络，实现冗余备份。

5. STP 的实现原理

交换机必须能够相互了解彼此之间的连接情况。为了让其他的交换机知道它的存在，每台交换机向网络中传送被称为 BPDU（Bridge Protocol Data Unit，网桥协议数据单元）的数据帧。

如果某台交换机能够从两条或多条链路上收到同一台交换机的 BPDU，则说明它们之间存在冗余链路，就会产生环路。

当存在环路时，交换机则使用生成树算法选择其中一条链路传递数据，并把某些相关的端口设置成阻塞（Blocking）状态以将它们的链路虚拟地断开，一旦当前正在使用的链路出现故障，就会把某个被阻塞的端口打开，使被阻断的链路接替原来的链路工作。

交换机之间通过 STP 交换信息，从而对环路进行定位并阻断多余链路，也就是通过封闭某些端口来消除网络中的环路，形成一个无环路的树形逻辑网络。

STP 的主要原理是当网络中存在冗余链路时，只允许主链路激活；当主链路因故障而中断时，激活冗余链路，保持网络连通。

6. 网桥协议数据单元（BPDU）

网桥或交换机为了让其他网桥或交换机知道它的存在，必须向端口传送被称为 BPDU（Bridge Protocol Data Unit，网桥协议数据单元）的数据帧。Catalyst 交换机每 10 分钟从所有的活动端口发出一个 BPDU。

7. 生成树算法（STA）

网桥接收到 BPDU 后，便利用 STA（Spanning Tree Algorithm，生成树算法）进行计算，就可以知道网络上是否存在环路。

当存在环路时，网桥就关闭冗余端口。关闭端口的过程被称为阻塞。被阻塞端口仍然是一个活动端口，即它仍然可以接收和读取 BPDU。这一过程一直持续到出现失败或者拓扑变化消除环路后才能结束。当这一过程结束后，端口便开始发送数据帧，因为这时环路已经不存在了。

8. 根桥（Root Bridge）

所有的树都有根，生成树也有根，它是一种特殊的网桥，并有一个非常恰当的名字：根桥。在网络中所有的网桥都被分配了一个称为优先级的数值，优先级的数值最小的网桥被称为根桥。

默认情况下，Catalyst 交换机的网桥优先级都为 32768，所以，如果只使用 Catalyst 交换机的话，所有网桥的优先级都一样。

在根桥的判定过程中要用到一种决胜法。

Catalyst 交换机地址池中最低优先级的 MAC 地址分配给监控机。这个 MAC 地址通常称为网桥 ID（BID）。将 MAC 地址作为 BID 可以确保有且只有一个 BID 值最小，因为 MAC 地址在全世界都是唯一的。

9. 根端口（Root Port）

所有非根网桥都要计算自己到根桥的距离，并选择距离根桥最近的端口为根端口。

数据通过链路是需要时间和带宽等开销的，开销的大小与链路的速率有关。端口到根桥的一条路径上的所有链路的累加开销就是端口到根桥的距离。根端口是非根网桥到根桥距离最近的那个端口。生成树链路开销值（Cost）如图 19-2 所示。

链路带宽	旧标准 Cost	新标准 Cost
4 Mbps	250	250
10 Mbps	100	100
16 Mbps	63	62
100 Mbps	10	19
155 Mbps	6	14
622 Mbps	2	6
1 Gbps	1	4
10 Gbps	1	2
>10 Gbps	1	1

图 19-2　生成树链路开销值（Cost）

10. 指定端口（Designated Port）

一个网段到根桥的所有路径中距离最短（开销最小）的直连端口被称为指定端口（Designated Port），其他所有路径中的直连端口就是非指定端口（Nondesignated Port）。

指定端口被标记为转发端口，能够转发数据帧，非指定端口则被阻塞不用。

11. 非指定端口（Nondesignated Port）

除根端口和指定端口之外的其他所有交换机端口都是非指定端口（Nondesignated Port）。

STP 最后阻塞全部非指定端口，从而使有环路的物理网络变成无环路的逻辑网络。该逻辑网络的拓扑结构就如一棵树，根桥就是它的树根，也是该逻辑网络的中心。

12. STP 的工作步骤

① 每个网络选举一个唯一的根桥。

One root bridge per network.

② 每个非根网桥选举一个唯一的根端口。

One root port per nonroot bridge.

③ 每个网段选举一个唯一的指定端口。

One designated port per segment.

④ 所有非指定端口都被阻塞不使用。

Nondesignated ports are unused.

13. STP 工作期间的端口状态转换

生成树经过一段时间（默认值是 50 秒左右）之后达到稳定状态，此时所有端口要么进入转发状态，要么进入阻塞状态。

14. 与 STP 有关的配置命令

打开 Spanning Tree 协议：

Switch(config)# spanning-tree

关闭 Spanning Tree 协议：

Switch(config)# no spanning-tree

注释：当交换机启动时 STP 是自动打开的。在模拟器中，默认是 PVST。

指定生成树协议模式（IEEE 802.1d）：

Switch(config)# spanning-tree mode stp

返回特权模式进行查看：

Switch# show spanning-tree

Switch# show spanning-tree vlan 1

配置交换机优先级（0 或 4096 的倍数，默认为 32768）：

Switch(config)# spanning-tree priority 8192

注释：数值越小优先级越高。

配置交换机端口优先级（0 或 16 的倍数，默认为 128）：

Switch(config-if)# spanning-tree port-priority 32

注释：数值越小优先级越高。

配置交换机端口的路径开销（1～20000000）：

Switch(config-if)# spanning-tree cost 1000

若要恢复到默认值，在相关命令前加 no。

15. 生成树协议调整

在连接单台主机的端口上使用下面的命令可加速生成树收敛时间：

spanning-tree portfast

在连接集线器、交换机和路由器等的端口上使用该命令会产生环路！

16. 思考

① 冗余拓扑结构解决了什么问题？
② 冗余拓扑结构带来了什么问题？
③ 生成树协议的作用是什么？
④ 在运行生成树协议的过程中交换机的端口要历经哪些状态？
⑤ 如何加速生成树的收敛过程？

17. 实训设备和环境

- 交换机：3 台（Catalyst WS 2950-24）；
- 带有网卡的 PC：3 台（带有超级终端）；
- 控制台电缆：3 条；
- 直连双绞线：3 条；
- 交叉双绞线：3 条；
- 用以上设备搭建网络拓扑，如图 19-1 所示；
- 模拟实训环境如图 19-3 所示。

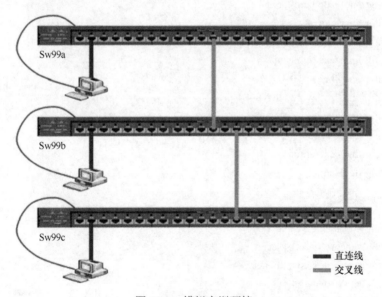

图 19-3　模拟实训环境

实训过程

1. 交换机基本配置

（1）Sw99a 基本配置

Switch(config)#hostname Sw99a

```
Sw99a(config)#enable secret 99secret
Sw99a(config)#service password-encr
Sw99a(config)#no ip domain-lookup
Sw99a(config)#line con 0
Sw99a(config-line)#logging sync
Sw99a(config-line)#exec-timeout 0 0
Sw99a(config-line)#line vty 0 4
Sw99a(config-line)#password 99vty0-4
Sw99a(config-line)#login
Sw99a(config-line)#logging sync
Sw99a(config-line)#exec-timeout 0 0
Sw99a(config-line)#end
```

（2）Sw99b 基本配置

```
Switch(config)#hostname Sw99b
Sw99b(config)#enable secret 99secret
Sw99b(config)#service password-encr
Sw99b(config)#no ip domain-lookup
Sw99b(config)#line con 0
Sw99b(config-line)#logging sync
Sw99b(config-line)#exec-timeout 0 0
Sw99b(config-line)#line vty 0 4
Sw99b(config-line)#password 99vty0-4
Sw99b(config-line)#login
Sw99b(config-line)#logging sync
Sw99b(config-line)#exec-timeout 0 0
Sw99b(config-line)#end
```

（3）Sw99c 基本配置

```
Switch(config)#hostname Sw99c
Sw99c(config)#enable secret 99secret
Sw99c(config)#service password-encr
Sw99c(config)#no ip domain-lookup
Sw99c(config)#line con 0
Sw99c(config-line)#logging sync
Sw99c(config-line)#exec-timeout 0 0
Sw99c(config-line)#line vty 0 4
Sw99c(config-line)#password 99vty0-4
Sw99c(config-line)#login
Sw99c(config-line)#logging sync
Sw99c(config-line)#exec-timeout 0 0
Sw99c(config-line)#end
```

2. PC 配置

在各台 PC 上配置 IP 地址、子网掩码和默认网关，然后单击【保存】按钮，保存已有配置。
① PC99a 的配置。
- IP 地址地址：10.99.1.254；
- 子网掩码：255.255.255.0；
- 默认网关：10.99.1.1。

② PC99b 的配置。
- IP 地址：10.99.1.253；
- 子网掩码：255.255.255.0；
- 默认网关：10.99.1.1。

③ PC99c 的配置。
- IP 地址：10.99.1.252；
- 子网掩码：255.255.255.0；
- 默认网关：10.99.1.1。

3. 网络连通性测试

（1）在 PC99a 上测试网络连通性

```
C:\>ping 10.99.1.253
!!!!!
C:\>ping 10.99.1.252
!!!!!
```

（2）在 PC99b 上测试网络连通性

```
C:\>ping 10.99.1.254
!!!!!
C:\>ping 10.99.1.252
!!!!!
```

（3）在 PC99c 上测试网络连通性

```
C:\>ping 10.99.1.254
!!!!!
C:\>ping 10.99.1.253
!!!!!
```

4. 生成树信息显示

（1）在 Sw99a 上显示生成树信息

```
Sw99a#show span
```

```
VLAN0001
  Spanning tree enabled protocol ieee
  Root ID    Priority    32769
             Address     0001.C77E.4EB8
             This bridge is the root                //根桥
             Hello Time   2 sec   Max Age 20 sec   Forward Delay 15 sec

  Bridge ID  Priority    32769  (priority 32768 sys-id-ext 1)
             Address     0001.C77E.4EB8
             Hello Time   2 sec   Max Age 20 sec   Forward Delay 15 sec
             Aging Time   20

Interface        Role Sts Cost      Prio.Nbr Type
---------------- ---- --- --------- -------- --------------------------------
Fa0/1            Desg FWD 19        128.1    P2p
Fa0/2            Desg FWD 19        128.2    P2p
Fa0/3            Desg FWD 19        128.3    P2p
//以上 3 行显示了指定端口的相关信息
```

（2）在 Sw99b 上显示生成树信息

```
Sw99b#show span
VLAN0001
  Spanning tree enabled protocol ieee
  Root ID    Priority    32769
             Address     0001.C77E.4EB8
             Cost        19
             Port        1(FastEthernet0/1)
             Hello Time   2 sec   Max Age 20 sec   Forward Delay 15 sec

  Bridge ID  Priority    32769  (priority 32768 sys-id-ext 1)
             Address     000A.F3C5.D5A1
             Hello Time   2 sec   Max Age 20 sec   Forward Delay 15 sec
             Aging Time   20

Interface        Role Sts Cost      Prio.Nbr Type
---------------- ---- --- --------- -------- --------------------------------
Fa0/3            Desg FWD 19        128.3    P2p
Fa0/2            Desg FWD 19        128.2    P2p
Fa0/1            Root FWD 19        128.1    P2p           //根端口
```

（3）在 Sw99c 上显示生成树信息

```
Sw99c#show span
VLAN0001
   Spanning tree enabled protocol ieee
   Root ID     Priority      32769
               Address       0001.C77E.4EB8
               Cost          19
               Port          2(FastEthernet0/2)
               Hello Time    2 sec   Max Age 20 sec   Forward Delay 15 sec

   Bridge ID   Priority      32769   (priority 32768 sys-id-ext 1)
               Address       0060.4764.6942
               Hello Time    2 sec   Max Age 20 sec   Forward Delay 15 sec
               Aging Time    20

Interface        Role Sts Cost       Prio.Nbr Type
---------------- ---- --- ---------  --------------------------------
Fa0/1            Desg FWD 19         128.1    P2p
Fa0/2            Root FWD 19         128.2    P2p
Fa0/3            Altn BLK 19         128.3    P2p       //阻塞端口 Fa0/3
```

生成树状态（一）如图 19-4 所示。

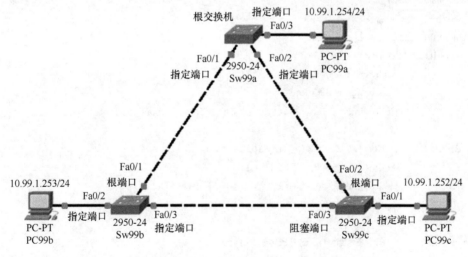

图 19-4　生成树状态（一）

（4）修改 Sw99b 的优先级

```
Sw99b(config)#span vlan 1 priority 16384      //这会使 Sw99b 被选为根桥
Sw99b(config)#exit
```

（5）再次在 Sw99a 上显示生成树信息

```
Sw99a#show span
VLAN0001
  Spanning tree enabled protocol ieee
  Root ID    Priority    16385
             Address     000A.F3C5.D5A1
             Cost        19
             Port        1(FastEthernet0/1)
             Hello Time   2 sec   Max Age 20 sec   Forward Delay 15 sec

  Bridge ID  Priority    32769   (priority 32768 sys-id-ext 1)
             Address     0001.C77E.4EB8
             Hello Time   2 sec   Max Age 20 sec   Forward Delay 15 sec
             Aging Time  20

Interface        Role Sts Cost      Prio.Nbr Type
---------------- ---- --- --------- --------------------------------
Fa0/1            Root FWD 19        128.1    P2p         //根端口
Fa0/2            Desg FWD 19        128.2    P2p
Fa0/3            Desg FWD 19        128.3    P2p
```

（6）再次在 Sw99b 上显示生成树信息

```
Sw99b#show span
VLAN0001
  Spanning tree enabled protocol ieee
  Root ID    Priority    16385
             Address     000A.F3C5.D5A1
             This bridge is the root        //根桥
             Hello Time   2 sec   Max Age 20 sec   Forward Delay 15 sec

  Bridge ID  Priority    16385   (priority 16384 sys-id-ext 1)
             Address     000A.F3C5.D5A1
             Hello Time   2 sec   Max Age 20 sec   Forward Delay 15 sec
             Aging Time  20

Interface        Role Sts Cost      Prio.Nbr Type
---------------- ---- --- --------- --------------------------------
Fa0/3            Desg FWD 19        128.3    P2p
Fa0/2            Desg FWD 19        128.2    P2p
Fa0/1            Desg FWD 19        128.1    P2p
//以上 3 行显示了指定端口的相关信息
```

（7）再次在 Sw99c 上显示生成树信息

```
Sw99c#show span
VLAN0001
  Spanning tree enabled protocol ieee
  Root ID    Priority      16385
             Address       000A.F3C5.D5A1
             Cost          19
             Port          3(FastEthernet0/3)
             Hello Time  2 sec   Max Age 20 sec   Forward Delay 15 sec

  Bridge ID  Priority      32769   (priority 32768 sys-id-ext 1)
             Address       0060.4764.6942
             Hello Time  2 sec   Max Age 20 sec   Forward Delay 15 sec
             Aging Time    20

Interface        Role Sts Cost      Prio.Nbr Type
---------------- ---- --- --------- -------- --------------------------------
Fa0/1            Desg FWD 19        128.1    P2p
Fa0/2            Altn BLK 19        128.2    P2p       //阻塞端口 Fa0/2
Fa0/3            Root FWD 19        128.3    P2p
```

生成树状态（二）如图 19-5 所示。

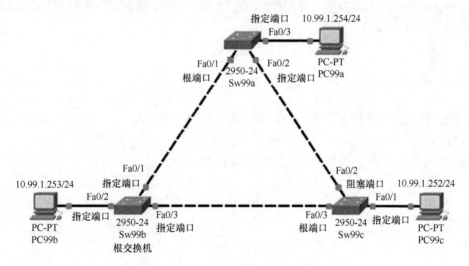

图 19-5　生成树状态（二）

如果将 Sw99c 的端口 Fa0/3 关闭，则新的生成树会是什么状态呢？
如果再将 Sw99c 的优先级修改为 8192，则新的生成树又会是什么状态呢？
请自行完成。

注意：STP 重新收敛时间较长，通常需要一分钟左右才会产生新的生成树。

保存配置：
- 验证配置正确后请保存配置，最好还能把配置文件备份到 TFTP 服务器上（如果有）；
- 在进行模拟实训时，不仅要保存配置，还要保存网络拓扑。

学习总结

为了增强网络的健壮性，我们经常会采用冗余拓扑，而这会产生交换环路。

交换环路会带来三个问题：广播风暴、多帧复制、交换机 CAM 表不稳定。

STP 用于消除交换环路，其基本原理是阻断一些交换机端口，构建一棵没有环路的生成树。

当网络发生故障时，STP 可以重新启用原来被阻断的端口，构建新的没有环路的生成树，自动调整网络的数据转发路径。

课后作业

完成上面的模拟实训，将实训过程的截图按顺序粘贴到一个 Word 文件里并用适当的文字说明你对它的理解；总结本次实训所需要的主要命令及其作用，作为实训报告上交。

实训报告一律以"ID 姓名项目号.doc"为文件名命名，网络拓扑及其配置也以"ID 姓名项目号.pkt"为文件名保存并上交。例如，张三的 ID 为 03，他的文件名为"03 张三 19.doc"和"03 张三 19.pkt"。

思考题

至此，我们已经学完了交换机的基本知识和基本配置。那么，如何将它们综合应用到网络构建的实际工作中呢？

在下一个项目中我们将学习有关的知识和技术。

项目 20　网络互联综合配置

项目描述

某单位需要构建一个由汇聚层和接入层构成的典型交换网络,并通过路由器接入因特网。若你是该单位的网络管理员,你将怎样构建该网络?

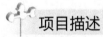

网络拓扑如图 20-1 所示。

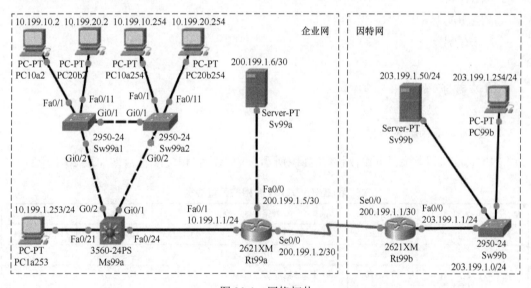

图 20-1　网络拓扑

网络拓扑图左边模拟一个企业网,右边模拟因特网。网络构建的总体目标是:使企业网内部全连通,使企业网能访问因特网,因特网只能访问企业网内的服务器 Sv99a。

学习目标

- 能规划和构建一个较复杂的网络;

- 熟练掌握路由器、三层交换机的配置；
- 掌握 NAT 的配置；
- 掌握路由配置和路由重新分布方法；
- 能模拟网络故障并掌握应急处理方法；
- 掌握对网络进行综合调试的技能；
- 知道怎么对网络进行测试和验收。

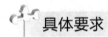

实训任务分解

① 路由器和各交换机基本配置。
② 路由器接口配置。
③ 交换机主干链路配置。
④ 交换机 VTP 配置。
⑤ 交换机 VLAN 配置。
⑥ 交换机 STP 配置。
⑦ NAT 配置。
⑧ 路由配置。
⑨ 保存配置。

具体要求

1．实训环境和配置数据

本项目实训环境如上面的网络拓扑图所示，VLAN 和相关配置数据如表 20-1 所示。

表 20-1　VLAN 和相关配置数据

交换机	VLAN ID	VLAN Name	端口	网络地址/掩码	设备	IP 地址	默认网关
Ms99a	1	default	Fa0/1~Fa0/24	10.199.1.0/24	PC1a253	10.199.1.253	10.199.1.254
Sw99a1	1	default	Fa0/21~Fa0/24	10.199.1.0/24			
	10	VLAN10	Fa0/1~Fa0/10	10.199.10.0/24	PC10a2	10.199.10.2	10.199.10.1
	20	VLAN 20	Fa0/11~Fa0/20	10.199.20.0/24	PC20b2	10.199.20.2	10.199.20.1
Sw99a2	1	default	Fa0/21~Fa0/24	10.199.2.0/24			
	10	VLAN10	Fa0/1~Fa0/10	10.199.10.0/24	PC10a254	10.199.10.254	10.199.10.1
	20	VLAN 20	Fa0/11~Fa0/20	10.199.20.0/24	PC20b254	10.199.20.254	10.199.20.1

2．具体要求

① 路由器和各交换机的基本配置：
- 主机名配置为拓扑图中所标示的主机名；

项目 20　网络互联综合配置

- 取消或禁止域名查找（即不会把输错的命令当作域名进行域名查找）；
- 进入特权模式的口令配置为 99（学号的最后两位数字）；
- 只能从 Console 端口登录（即禁止远程登录），而且登录时无须输入口令；
- 登录后不会超时退出（即很久无任何操作也不会自动断开会话连接）；
- 系统信息同步（当输入命令时未按回车前不会自动显示系统信息）。

② 交换机 Ms99a、Sw99a1 和 Sw99a2 构成 VTP 域 a99，VTP 口令配置为 99aaa，将 Ms99a 配置为 VTP Server，Sw99a1 和 Sw99a2 都配置为 VTP Client。

③ 企业网内的 VLAN 1、VLAN 10 和 VLAN 20 的根桥必须确保都是 Ms99a。

④ 串行线路要配置 PPP 封装，Rt99b 为 DCE 端，时钟频率为 2000000 Hz。

⑤ 为了使企业网能与因特网通信，该企业申请获得的注册 IP 地址为 200.199.1.0/29，其中一半即 200.199.1.0/30 已用于与因特网的串行连接，另一半即 200.199.1.4/30 用于服务器 Sv99a 直接上因特网（无须进行 NAT），使因特网的主机都能访问它。虽然因特网的主机不能访问企业网的其他主机，但企业网的其他主机都要能访问因特网，即要为企业网内的其他主机配置 NAT（实为 PAT）功能。

⑥ 路由配置：全部采用静态路由，且要尽量减少路由条目以达到最佳配置，还要注意不能将企业网内部的私有 IP 地址泄露到因特网上。

⑦ 为了实现网络建设的总体目标，没有明示或暗示的其他配置可灵活处理。

实训过程

完成实训的每一步后都应将相关配置保存到一个文件中。例如，完成网络拓扑搭建后，将其保存为文件名为 P20001.pkt 的文件，然后再将其另存为文件名为 P20002.pkt 的文件；完成基本配置后，将相关配置保存到文件 P20002.pkt 中，然后再将其另存为文件名为 P20003.pkt 的文件；完成接口配置后……

1. 路由器和各交换机基本配置

（1）路由器 Rt99a 基本配置

```
Router(config)# host Rt99a
Rt99a(config)# no ip domain-lookup
Rt99a(config)# enable secret 99
Rt99a(config)# line con 0
Rt99a(config-line)# logg sync
Rt99a(config-line)# exec-timeout 0 0
Rt99a(config-line)# exit
```

（2）路由器 Rt99b 基本配置

```
Router(config)# host Rt99b
Rt99b(config)# no ip domain-lookup
```

```
Rt99b(config)# enable secret 99
Rt99b(config)# line con 0
Rt99b(config-line)# logg sync
Rt99b(config-line)# exec-timeout 0 0
Rt99b(config-line)# exit
```

（3）交换机 Ms99a 基本配置

```
Switch(config)# host Ms99a
Ms99a(config)# no ip domain-lookup
Ms99a(config)# enable secret 99
Ms99a(config)# line con 0
Ms99a(config-line)# logg sync
Ms99a(config-line)# exec-timeout 0 0
Ms99a(config-line)# exit
```

（4）交换机 Sw99a1 基本配置

```
Switch(config)# host Sw99a1
Sw99a1(config)# no ip domain-lookup
Sw99a1(config)# enable secret 99
Sw99a1(config)# line con 0
Sw99a1(config-line)# logg sync
Sw99a1(config-line)# exec-timeout 0 0
Sw99a1(config-line)# exit
```

（5）交换机 Sw99a2 基本配置

```
Switch(config)# host Sw99a2
Sw99a2(config)# no ip domain-lookup
Sw99a2(config)# enable secret 99
Sw99a2(config)# line con 0
Sw99a2(config-line)# logg sync
Sw99a2(config-line)# exec-timeout 0 0
Sw99a2(config-line)# exit
```

2. 路由器接口配置

（1）配置路由器 Rt99a 的接口

```
Rt99a(config)# int fa0/0
Rt99a(config-if)# desc LAN link to Sv99a
Rt99a(config-if)# ip addr 200.199.1.5 255.255.255.252
Rt99a(config-if)# no shut
Rt99a(config-if)# int fa0/1
Rt99a(config-if)# desc LAN link to Ms99a
Rt99a(config-if)# ip addr 10.199.1.1 255.255.255.0
Rt99a(config-if)# no shut
```

```
Rt99a(config-if)# int se0/0
Rt99a(config-if)# desc WAN link to Rt99b
Rt99a(config-if)# ip addr 200.199.1.2 255.255.255.252
Rt99a(config-if)# encap ppp
Rt99a(config-if)# no shut
Rt99a(config-if)# exit
```

（2）配置路由器 Rt99b 的接口

```
Rt99b(config)# int fa0/0
Rt99b(config-if)# desc Lan link to Sw99b
Rt99b(config-if)# ip addr 203.199.1.1 255.255.255.0
Rt99b(config-if)# no shut
Rt99b(config-if)# int se0/0
Rt99b(config-if)# desc WAN link to Rt99a
Rt99b(config-if)# ip addr 200.199.1.1 255.255.255.252
Rt99b(config-if)# encap ppp
Rt99b(config-if)# clock rate 2000000
Rt99b(config-if)# no shut
Rt99b(config-if)# exit
```

3. 交换机主干链路的配置

（1）在交换机 Ms99a 上配置主干链路

```
Ms99a(config)# int gi0/1
Ms99a(config-if)# switch trunk encap dot1q
Ms99a(config-if)# switch mode trunk
Ms99a(config-if)# int gi0/2
Ms99a(config-if)# switch trunk encap dot1q
Ms99a(config-if)# switch mode trunk
Ms99a(config-if)# exit
```

（2）在交换机 Sw99a1 上配置主干链路

```
Sw99a1(config)# int gi0/1
Sw99a1(config-if)# switch mode trunk
Sw99a1(config)# int gi0/2
Sw99a1(config-if)# switch mode trunk
Sw99a1(config-if)# exit
```

（3）在交换机 Sw99a2 上配置主干链路

```
Sw99a2(config)# int gi0/1
Sw99a2(config-if)# switch mode trunk
Sw99a2(config)# int gi0/2
Sw99a2(config-if)# switch mode trunk
```

Sw99a2(config-if)# exit

4. 交换机 VTP 配置

（1）在交换机 Ms99a 上配置 VTP

Ms99a(config)# vtp version 2
Ms99a(config)# vtp domain a99
Ms99a(config)# vtp mode server
Ms99a(config)# vtp password 99aaa

（2）在交换机 Sw99a1 上配置 VTP

Sw99a1(config)# vtp version 2
Sw99a1(config)# vtp domain a99
Sw99a1(config)# vtp mode client
Sw99a1(config)# vtp password 99aaa

（3）在交换机 Sw99a2 上配置 VTP

Sw99a2(config)# vtp version 2
Sw99a2(config)# vtp domain a99
Sw99a2(config)# vtp mode client
Sw99a2(config)# vtp password 99aaa

5. 交换机 VLAN 配置

（1）在交换机 Ms99a 上配置 VLAN

Ms99a(config)# vlan 10
Ms99a(config-vlan)# name VLAN10
Ms99a(config-vlan)# vlan 20
Ms99a(config-vlan)# name VLAN20
Ms99a(config-vlan)# exit
Ms99a(config)# int vlan 1
Ms99a(config-if)# ip addr 10.199.1.254 255.255.255.0
Ms99a(config-if)# no shut
Ms99a(config-if)# int vlan 10
Ms99a(config-if)# ip addr 10.199.10.1   255.255.255.0
Ms99a(config-if)# int vlan 20
Ms99a(config-if)# ip addr 10.199.20.1   255.255.255.0
Ms99a(config-if)# exit

（2）在交换机 Sw99a1 上配置 VLAN

Sw99a1(config)# int range fa0/1-10
Sw99a1(config-if-range)# switch access vlan 10
Sw99a1(config-if-range)# exit

```
Sw99a1(config)# int range fa0/11-20
Sw99a1(config-if-range)# switch access vlan 20
Sw99a1(config-if-range)# end
```

（3）在交换机 Sw99a2 上配置 VLAN

```
Sw99a2(config)# int range fa0/1-10
Sw99a2(config-if-range)# switch access vlan 10
Sw99a2(config-if-range)# exit
Sw99a2(config)# int range fa0/11-20
Sw99a2(config-if-range)# switch access vlan 20
Sw99a2(config-if-range)# end
```

（4）在各交换机上查看 VLAN 信息以验证上面配置的正确性

```
Ms99a# show vlan brief
Sw99a1# show vlan brief
Sw99a2# show vlan brief
```

6. 交换机 STP 配置

（1）在交换机 Ms99a 上配置 STP

```
Ms99a(config)# span vlan 1 priority 4096
Ms99a(config)# span vlan 10 priority 4096
Ms99a(config)# span vlan 20 priority 4096
Ms99a(config)# end
```

（2）在交换机 Ms99a 上查看 STP 信息以验证各 VLAN 的根桥都是 Ms99a

```
Ms99a# show span vlan 1
Ms99a# show span vlan 10
Ms99a# show span vlan 20
```

7. NAT 配置

在路由器 Rt99a 上配置 PAT 功能：

```
Rt99a(config)# access-list 2 permit 10.199.0.0 0.0.255.255
Rt99a(config)# ip nat pool NP1 200.199.1.2 200.199.1.2 netmask 255.255.255.252
Rt99a(config)# ip nat inside source list 2 pool NP1 overload
Rt99a(config)# int fa0/1
Rt99a(config-if)# ip nat inside
Rt99a(config-if)# int se0/0
Rt99a(config-if)# ip nat outside
Rt99a(config-if)# end
```

8. 路由配置

（1）在路由器 Rt99a 上配置静态路由

Rt99a(config)# ip route 0.0.0.0 0.0.0.0 200.199.1.1
Rt99a(config)# ip route 10.199.0.0 255.255.0.0 10.199.1.254
Rt99a(config)# end

（2）在路由器 Rt99b 上配置静态路由

Rt99b(config)# ip route 200.199.1.0 255.255.255.248 200.199.1.2
Rt99b(config)# end

（3）在交换机 Ms99a 上配置默认路由

Ms99a(config)# ip routing
Ms99a(config)# ip route 0.0.0.0 0.0.0.0 10.199.1.1
Ms99a(config)# end

（4）查看路由信息，验证路由配置的正确性

Rt99a# show ip route
Rt99b# show ip route
Ms99a# show ip route

9. 保存配置

在各设备上保存配置数据：

Rt99a# write
Rt99b# write
Ms99a# write
Sw99a1# write
Sw99a2# write

判断配置正确与否最可靠的方法是测试。关闭 Ms99a 的接口 Gi0/1 或 Gi0/2，等网络稳定后，在 PC1a253 上 ping 10.199.20.2 或 ping 10.199.20.254，都能通；ping 两台服务器的 IP 地址也都能通。请自行全面测试网络的连通性。

最后，请保存文件。如果忘了保存，可重做一次。最好多做几次。

学习总结

至此，我们已经学完了路由器和交换机的一些基本知识和各种常用配置。在本项目中，我们综合应用了以前所学的各种配置，希望通过思考都能深刻理解和熟练应用。

在实际建网时，还要考虑高效性、安全性、可靠性和便于维护等一系列问题。

在管理网络时，还要学习网络故障诊断、调试和解决方法。

这些更复杂的内容将在进阶篇中介绍和学习。

项目 20　网络互联综合配置

课后作业

完成上面的模拟实训，将实训过程的截图按顺序粘贴到一个 Word 文件里并用适当的文字说明你对它的理解；总结本次实训所需要的主要命令及其作用，作为实训报告上交。

实训报告一律以"ID 姓名项目号.doc"为文件名命名，网络拓扑及其配置也以"ID 姓名项目号.pkt"为文件名保存并上交。例如，张三的 ID 为 03，他的文件名为"03 张三 20.doc"和"03 张三 20.pkt"。

思考题

该网络还有哪些不足或还可增加什么配置？

如何进一步提高企业网的安全性？